A BOOK OF TIME

A BOOK OF TIME

by Lesley Coleman

LONGMAN

LONGMAN GROUP LIMITED
LONDON

Associated companies, branches and representatives
throughout the world

© Lesley Coleman 1971

First published 1971

ISBN 0 582 15031 0

Printed in Great Britain by

Lowe & Brydone (Printers) Ltd London NW10.

CONTENTS

Introduction

In the year 1519 Ferdinand Magellan, the Portuguese explorer, sailed from Spain with a fleet of five ships and 270 men to find a westward route to the Spice Islands, where the profitable spices came from off the coast of Asia. He sailed down round the southernmost tip of South America, emerged into a vast ocean so calm that he called it the Pacific, and then, with little idea of where he was, sailed on west. His crews mutinied and starved, ate sawdust, boot leather, and any ships' rats they could catch. Then, after three months without sight of land, they came first to the Philippines and at last to the Spice Islands. Still sailing west, they came home to Spain.

Magellan had been killed on the way; of the five

ships that set out only one remained, and only nineteen men survived. They were the first men ever to sail round the world. Yet when they got home they were made to do penance because on their three years' voyage they had lost count of time, had confused their calendars, and had unknowingly committed the sin of celebrating some saints' festivals on the wrong days.

It is not surprising that they lost themselves and did not know exactly where they were or what day it was. Nor is it surprising that the church was concerned to make them do proper penance for confusing their saints' days, because the making and keeping of calendars—which are the ways of numbering the months and days of the year—had always been priests' work. And, in the past, the sequence of festivals was a secret to be revealed to men only by priests, whose business it was to interpret the wishes of gods.

Perhaps Stonehenge was a calendar. It was certainly a temple. Between 1800 and 1400 B.C., men dragged stones weighing as much as 100 tons to Salisbury Plain from as far as 140 miles away. There they arranged them so that on Midsummer day, the longest day of the year, the sun rose directly over the heel stone and shone down an avenue to the altar stone at the centre. Even our present calendar is the work of a priest, Pope Gregory XIII, who, about 1580, was concerned to arrange matters so that the Christian festival called Easter should come always in the spring: he caused the Gregorian calendar to be devised.

Calendars are more than pieces of paper with sheets to tear off for months. They are, and always have been, sometimes the natural, but more often the religious or political, arrangements of the seasons and festivals and days of the year, and they were not easily arrived at.

Nor were clocks easily made. The very cheapest tin clock you could buy anywhere today, would have been a gift to Magellan. His sailors lost count of time because they had no clocks at all, or anything by which to measure time accurately. They navigated by chance and instinct, and since they did not know *when* they were, they could not, by observing the stars and making an exact calculation of time and distance, know *where* they were.

And what is a clock? Now it is a machine that has hands, and a dial, and usually a tick. The earliest mechanical clocks had none of these things. And many clocks were not machines at all. Carl Linnaeus, the Swedish naturalist, made a flower clock by planting round a border certain flowers which he knew opened or closed at an exact time each day during the time when they flower. The Star of Bethlehem opens at 11 in the morning, the Scarlet Pimpernel closes at two in the afternoon, and the Evening Primrose opens at six. This was in the eighteenth century, when it was a toy. But in the fifteenth century it would have been more accurate than most moving, iron clocks. Until well on into the seventeenth century, sundials were habitually used to correct erratic mechanical clocks.

So a calendar is a way of counting days and months, and a way of organising society. A clock is a way of accurately measuring the moments as they pass. Months, days, and moments make time. But Time, with a capital T, is also a complex idea. And in these days, with men going to the moon, Time is an exact science. We have come on a bit. No men in space will be so lost as Magellan's nineteen survivors.

This book then is about time and about men: not only those who have worked with clocks and calendars to make them more exact but those who have worked with Time in every sense of that apparently simple word.

Chapter 1   Early Calendars

J. B. Priestley describes calendar-making as 'a long struggle to make a tidy job out of the rather untidy natural units of time-measurement'. From very earliest times, especially after people gave up the nomadic life and settled down to planting and growing, they needed calendars. They had to know in advance when the time would be right to prepare the ground for planting or to prepare themselves for the harshness of winter. And because they believed that good weather and good harvests depended on the humour of gods who controlled such things, they had to know when it was fitting to hold festivals in honour of those who could give them what they wanted. By the rising and setting of the sun they knew days; the waxing and waning of the moon gave them months. Plants had times for sowing, growing, reaping and withering according to the position of the earth in its journey round the sun; and so people knew the seasons which make up a year. But the sun and the moon, 'the natural units of time measurement', did not fit easily together into a regular predictable year.

The priests of the Sumerians who flourished in Lower Mesopotamia in about 3000 B.C. are the first known calendar-makers to make a reasonably 'tidy job' of it.

It was the priests who were responsible for making calendars because they were also

Tablet in Sumerian describing the rebuilding by King Hammurabi of a temple to the sun god c. 1760 B.C. (British Museum)

astronomers who observed the movements of the heavenly bodies and could interpret from them the moods and wishes of the gods. Their calculations, which would have been set down on clay tablets, are lost, but we know something about their methods from the records of the Babylonians who followed the Sumerians as rulers of Mesopotamia.

It seems that the Sumerians based their calendar chiefly on the waxing and waning of the moon. They observed that there are about twenty-nine or thirty days from one new moon to the next and that after about twelve new moons the cycle of the seasons began to repeat itself. So they divided the year into twelve months of thirty days each. But this simple scheme alone was not satisfactory. The problems which face all calendar-makers are that the earth takes almost $365\frac{1}{4}$ days to complete its journey around the sun, and that from one new moon to the next there are about twenty-nine and a half days. Twelve months of thirty days each do not fit either the sun or the moon accurately. They give a short year of 360 days. The months gradually get out of step with the seasons and finally give men no guide to planting or ploughing times. The fourth month of a year based entirely on the moon may be in spring to begin with, the time of planting and of praising the god who can make things grow, but in less than forty years it will fall in midwinter.

Though we do not know how the Sumerians overcame this difficulty we do know what the Babylonians did. From their very precise observations the astronomer-priests made maps and time-tables of the stars' positions on clay tablets and on these they based the calendars which regulated crop-growing and religious ceremonies. They had a year of twelve months which were alternately twenty-nine and thirty days long to keep them in step with the twenty-nine and a half days of one lunation. It was a year of only 354 days, still a long way short of the solar year of $365\frac{1}{4}$ days. So the Babylonians simply added, or intercalated, an extra month before the twelfth month of the year every three years or so. It is recorded in a letter written about 1700 B.C. that their founder, King Hammurabi, ordered an extra month to be added whenever he was told by the priests that the year had 'a deficiency'.

By 432 B.C. it had been discovered that after a period of nineteen years the apparent movements of the moon and sun fall into step, that the full and new moons come at roughly the same time in the corresponding solar years of a 19-year cycle. This discovery made possible a more systematic method of intercalation. The Babylonians found that they

could correct their calendar with its years of 354
days by adding an extra month to seven particular
years of the nineteen-year cycle. There was an error
of one day in this scheme which was put right by a
man called Calippus in 330 B.C. when he laid down
that at the end of four cycles, seventy-six years, one
day should be omitted from the year. We now call
these nineteen years the Metonic Cycle after an
Athenian astronomer called Meton who either
worked it out for himself or brought the knowledge
of it to Greece from Babylon.

The Babylonians kept a seven-day week because
each day was devoted to the worship of a different

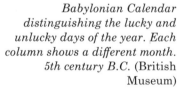

*Babylonian Calendar
distinguishing the lucky and
unlucky days of the year. Each
column shows a different month.
5th century B.C.* (British
Museum)

heavenly body. Five planets were known to them: Mercury, Venus, Mars, Jupiter and Saturn, which were worshipped together with the sun and the moon. Thus, though the length of a month and of a year were roughly governed by the moon and the sun, the week had seven days simply by chance. In later civilisations the length of the week varied from four to ten days. It was governed by the frequency of markets. The now universal seven-day week came in with the spread of Christianity and Mohammedanism. It derives from the story of the creation; God worked six days making the world and rested on the seventh. But the Jews and Mohammedans probably borrowed it from the Babylonian calendar which was the model for the Hebrew and Moslem calendars which are still used today.

Mohammed forbade the intercalation of months, so the Moslems, at least in their religious observances, keep a purely lunar calendar which has no relation at all to the solar year or to the seasons.

Christians date their years according to the birth of Christ and so put B.C., before Christ, to indicate the years before he was born and A.D., an abbreviation of the Latin Anno Domini, meaning the year of our Lord, to indicate the years since his birth. The Moslem calendar starts from the flight

or Hejira of Mohammed to Medina. It happened in 622 A.D. according to the Christian calendar. So, taking into account the short Moslem year, 1968–1969 A.D. was in the Moslem calendar A.H. (Anno Hejiræ) 1388–1389. Many countries like Persia, Turkey and Arabia still use the Moslem calendar privately, though they may use ours, which we call the Gregorian calendar, in public.

The Jews correct their lunar calendar as the Babylonians did by adding an extra month to seven years of the nineteen-year Metonic Cycle. These embolismic years, as they are called, are the third, sixth, eighth, eleventh, fourteenth, seventeenth and nineteenth of the cycle. The starting point of the Jewish calendar is the creation of the world which happened, the Jews believe, in October 3761 B.C.

So A.D. 1969 was A.M. (Anno Mundi or Year of the World) 5730 in the Jewish calendar.

Nowadays we use a calendar which is based on that of the Egyptians and which came to us through the Romans. The Early Egyptians who flourished at the same time as the Sumerians were also, like them, efficient calendar-makers.

The most important event in the Egyptian year was the flooding of the Nile, because the Egyptians depended on the floods to water and fertilize their land ready for crop-planting. It so happened that Sirius, the Dog Star, generally had its heliacal rising just before the floods began. The heliacal rising of a star is when it emerges once again from the sun's rays and appears above the horizon just before sunrise. For some time before this it has been above the horizon only after sunrise and so it is not visible, since of course the brightness of the sun kills the light of the stars. It seemed to the Egyptians that Sirius, or Sothis, the Egyptian name for the Dog Star, was returning to the sky after a long absence and its return brought the longed-for floods. So, in about 3000 B.C., the priests made the rising of Sothis the start of the New Year. There were apparently about 365 days between one heliacal rising and the next and so that they would accord with the movements of the moon-god, who was still important, the Egyptians simply added five days regularly to the end of a year composed of twelve months of thirty days each. Those five extra days were called 'Days of the Year' and were holidays devoted to religious festivals in honour of Sothis.

This calendar was used until 238 B.C. But 365 days are about six hours shorter than it takes from one heliacal rising of Sirius to the next, so very slowly the calendar got out of step with the seasons. Ptolemy III, one of the rulers of Egypt, proposed that the error should be put right by observing a sixth 'day of the year' every four years. On the Canopus Stone, which was set up in 238 B.C. in honour of Ptolemy and his consort Berenice, it says he decreed that 'from this time onwards one day as festival of the good-doing gods shall be added every four years to the five additional days which come before the new year, so that it may happen that every man shall know the small amount (of time) which was lacking in the arrangement of the seasons, and of the year, and the things which passed as law for the knowledge of their movements hath been corrected'. But by now, since it had become accepted that the religious feasts moved through the seasons and the festival of Sothis was not necessarily held at the time of his heliacal rising, the priests refused to add this extra day every four years. Two separate calendars were therefore used, a religious one with moving festivals and one for farmers who could not ignore the seasons. Ptolemy's scheme went unaccepted. Almost 200 years later it was adopted by the Romans under Julius Cæsar.

In the earliest Roman calendar there were only ten lunar months beginning with March. It was an agricultural calendar and so about sixty days of winter were simply omitted from the calculations because there is very little work to be done in the fields at that time. Numa Pompilius, who traditionally was the second king of Rome from 715–672 B.C., is said to have reformed the calendar by adding two more months to the end of the year. And he assigned dates to the many religious festivals and entrusted their due observation to the priests.

The religious calendar was a lunar one of 355 days and it was obvious that it did not fit the solar year. The priests were responsible for intercalating days or months whenever necessary. It was a responsibility which they frequently abused. Corrupt priests could arrange things to suit their own ends. By intercalating or neglecting to do so they might lengthen or shorten the term of office of a friend or enemy, postpone or hasten the day for an important trial, lengthen or shorten the period of a contract.

Some of the earliest information we have about ancient Rome comes from the priests' calendars of festivals and feasts written on stone. Like those of Sumeria and Egypt the priests of Rome were extremely powerful. On the calends (the latin word which has given us the word calendar), the time of the new moon and so the first day of a month, the Roman people assembled to learn from them the programme of feasts for the coming month. There would also be a certain number of 'nefasti' (days made unlucky by the threatening position of the stars). On feast days and unlucky days no business was allowed. There were many festivals. The days of full moon were sacred to Jupiter; wine festivals were held in his honour too. A great warrior festival in March honoured Mars, the god of war. Ceres, the goddess of growth and germination, was celebrated in April.

There was some confusion about the start of the year. The priests' New Year began in March with the coming of spring, but from 153 B.C. the consuls began their year of office on January 1. In 153 B.C. Rome was at war with Spain and new consuls had taken over on January 1 as an emergency measure which like so many emergency measures became an accepted custom. Finally, though the priests might ignore the solar year, Roman farmers, like those in Egypt, were bound by it, so they observed a Year of Seasons fixed by the sun and beginning, like that of the priests, with March and the spring.

The Romans dated their calendar from the founding of Rome, 'Ab Urbe Condita', by Romulus and Remus. 1 A.U.C. is the equivalent of about 753 B.C.

Forty-six years before the birth of Christ, when Julius Caesar was at the height of his power he found that the calendar for the civil year was nearly two months behind that of the solar year. Thus all the religious ceremonies came at the wrong time of year; the festival of Flora, a spring ceremony which should have been celebrated in April, was now observed in July. Some sense had to be brought back to the calendar and it also had to be freed from the power of unreliable priests. Cæsar called in astronomers and mathematicians to help him plan a reform. Among them was a Greek astronomer, Sosigenes, who is thought to have come originally from Alexandria in Egypt.

It was decided that the year should from now on officially start on January 1. Not only was this the day when new consuls took office but also, in the first year of the reform, 45 B.C., it was the date of

Julius Cæsar as he appears on a Roman coin. (British Museum)

the first full moon after the shortest day of 46 B.C. By honouring the moon in this way Cæsar pleased the priests and religious people who still considered the moon more important than the sun.

The lunar year had to be increased by ten days. The extra days were divided up between the months so that odd-numbered months, January, March, and so on, had thirty-one days, while the even-numbered months had thirty days. February was the exception with twenty-nine days ordinarily and thirty only in a leap year. To cope with the odd quarter day of the sun's yearly cycle an extra day was put in every four years to make what we now call a leap year; this was the reform Ptolemy III had advocated two centuries earlier. The Romans called our leap year the bissextile year, meaning literally the second six year. This was because they added the extra day after February 24 or, in their way of reckoning, after the sixth day before the calends (first) of March; the extra day was seen as a repetition of the sixth day, a second sixth, before the calends. Between February 24 and 25 had long been a traditional time for intercalation. We call the bissextile year a leap year because after it a fixed festival comes two days later than it has done in the previous year, thus seeming to leap over a day. So if someone's birthday was on a Tuesday in May 1967 it fell on a Thursday in 1968 because 1968 was a leap year; Wednesday had been 'leapt' over.

After the new calendar was published Cæsar declared it binding on the State. It came into use from the calends of January 45 B.C. Writers sometimes call 46 B.C. the Year of Confusion because it had to be 445 days long to make the new year and the new calendar start right. Some Romans, like Cicero, the writer, who were hostile to Caesar because of the great power he wielded, saw his calendar reform simply as a way for him to lengthen his term of office.

Yet the Julian calendar was used by all civilized Europe almost unchanged until 1582. The Christian world even continued to number the years from the founding of Rome until A.D. 527. In that year an abbot of Rome, Dionysius Exiguus, introduced to

24

Italy our present method of dating before and after Christ's birth, which we call the Christian era or epoch. It is now generally accepted that he miscalculated the date of Christ's birth. Historical evidence suggests that he was born several years earlier than Dionysius Exiguus estimated, most probably in 4 B.C. St Augustine is said to have brought the system to England in A.D. 596 but as bishops at the Council of Chelsea ordered its use in 816 A.D. it had clearly not been generally adopted by that time.

Dionysius Exiguus was also responsible for reintroducing a muddle which Julius Cæsar had cleared up. He established that the year should start on March 25 rather than on January 1. And in fact, until 1782, the two systems were used in England. The legal year began in March and the historical year in January.

Modern Calendar Reforms Chapter 2

The Julian Calendar is based on the assumption that the earth takes exactly 365 days and 6 hours to complete one journey around the sun. But really it takes 365 days, 5 hours, 48 minutes, 46 seconds. Caesar's calendar reformers allowed an apparently trifling error but clearly over hundreds of years the seconds build up into minutes, then hours, then days.

At the Council of Nicea in A.D. 325 important members of the Christian church had, after long discussion, agreed to a regular way of dating Easter. From that time on it was to come on the first Sunday after the first full moon after March 21. In the year of the Council of Nicea, March 21 was the date of the vernal equinox which is officially

Pope Gregory XIII with his committee of churchmen and scientists debating the new calendar. 16th century painting in the Archivio di Stato, Siena.
(Foto Grassi, Siena)

the first day of spring. Vernal means spring and an equinox is that time of the year (there is another in autumn), when day and night are of equal length.

By the sixteenth century the Roman Catholic church was worried again about the dating of Easter. The Julian Calendar's slight error of seconds in the length of the year had accumulated so that March 21, the official date used for fixing

Easter, was no longer the day of the vernal equinox, the important event on which the time of Easter depends. In fact the vernal equinox came at that time about ten days before March 21. If the error was allowed to continue Easter would come to be celebrated in summer and later on in winter according to the sun; apart from anything else this would make nonsense of the symbolism of Easter as a time of renewal and rebirth.

As early as the thirteenth century Roger Bacon, a Franciscan friar, scientist and philosopher, noticed the error in the Julian calendar and wrote a treatise advocating reform which he sent to the Pope. Nothing was done. By the fifteenth century the difference between the vernal equinox and its calendar date of March 21 amounted to nine days. So in 1472 Pope Sixtus IV invited Johann Müeller, an astronomer, to Rome to supervise changes in the calendar. Müeller was assassinated in 1476. The calendar remained unchanged.

Then in the sixteenth century Pope Gregory XIII called upon Aloysius Lilius, an astronomer and professor of medicine at the University of Perugia, for his advice and help in reforming the calendar. He spent ten years studying the matter but he died in 1576 without completing the work. A German Jesuit and mathematician, Christopher Clavius, verified Lilius's calculations and worked out the rules for the change. He wrote two long books in Latin defending the new calendar which helped persuade many doubters to accept it. A committee of clergymen and scientists appointed by the Pope debated the reform and it was accepted by Gregory.

By Papal decree, 1582 was shortened by ten days; October 4 was followed immediately by October 15. And there was a revision of the leap year rule. According to the Julian calendar every year which was exactly divisible by four was a leap year. Gregory decreed that the last year of a century should not be a leap year unless its number was divisible by 400. So for example the years 1800 and 1900 were not leap years. In this way three leap years are omitted every four centuries. These new calculations make the Gregorian calendar fit the solar calendar accurately to within one day every 3,323 years, an amount so small that no one has yet found it necessary to make any adjustments. Gregory also decided, once again, that New Year's Day should be on January 1.

The reformed calendar was quickly adopted by Roman Catholic states and countries, like Spain, Portugal and France. Protestants were slower to accept a scheme advocated by the Roman Catholic church. But, as a matter of convenience, the Protestant countries did gradually come to accept the Gregorian calendar. England managed to withstand change until 1752 when she was eleven days out of step with the greater part of Europe.

For some time before 1752 writers in the *Gentleman's Magazine*, which was a monthly magazine published in London, had favoured some kind of calendar reform. At first it seems that

Frontispiece of the Gentleman's Magazine for 1752. The man crowned with a laurel wreath seems to be the editor supported by Old Father Time (see page 132) who is presenting the new Calendar to an agreeable Boadicea, representing Great Britain.

they were less concerned with being at odds with neighbouring countries than they were that there should be one generally accepted date in England for starting the New Year. In January 1735 a writer complained that all lawyers dated the new year from March 25. Some newspapers did the same, but some started the year in January. Others, uncertainly, put the date of the two possible years, 1734–1735 for example, until March 25 had passed. 'The monstrous Absurdity of such a Variety of Dates is too notorious to every judicious Reader to be denied.' This choice of dates was felt to be confusing not only for the people of the time but also for future historians who would, the writer feared, have great difficulty in sorting out the exact dates of events. Our newspapers were 'the only Papers in the World that have not any certain Dates to show when they were published'.

A correspondent who signed himself T. W. Wisbech, wrote, 'I wish that our Calendar and our Almanacks were so far reform'd as to observe the Year of Julius Cæsar exactly without any Alteration. We should not then have two different beginnings to the same Year, January and March . . . certainly a ridiculous Custom.' He also wanted the extra day of leap year to be put after the 24th of February as Cæsar had arranged it because 'to intercalate this Day at the end of the Month is not only defective in grammar but sense. February 28 is call'd in the Roman Calendar "pridie calendarum Mars" (the day before the calends of March) which name alone sufficiently testifies that no Day can intervene between that and the Calends of March.' It seems a pedantic point since probably no one in England ever called February 28 'pridie calendarum Mars'. In answer to this letter another correspondent said that it was at the Restoration of Charles II that the feast of St Matthias was set on February 24 and the leap year day was then put at the end of the month, 'its being so placed by the Reformers of our Calendar and Liturgy . . . in solemn Synod lawfully convened'. Leap Year day was put at the end of the month rather than where Caesar had put it so that there could be no confusion about when to celebrate the feast of St Matthias. This same writer, Edmund Weaver of Friestone, wanted a change in the calendar because 'in Process of Time our Christmas will be kept at Midsummer; Easter in the Place of St Michael; and Whitsuntide will be slipped into the Winter Quarter'. In December 1744 he had support from Tho. Whiston; 'We want another Council to assemble now for the same reason as the Nicene did, if for nothing else, at least to regulate the true time of Easter.' This correspondent did not, it seems, contemplate the easiest solution which was to adopt the Gregorian calendar used already by most European countries. But, by March 1747, Will Chapple of Exon got around to the suggestion that eleven days should be omitted from the calendar as used in England and said: 'Let the reckoning be continued as in the Gregorian account omitting three leap years in 400 years.' He even had a way to rectify the small error in the Gregorian calendar: 'In the year of our Lord 6,000 (if the world continue so long) let the leap day which is retained in the Gregorian account be omitted.' In a postscript he said that there was talk of calendar reform during the 'present session of Parliament'.

But still it was not until February 1751 that a Bill for altering 'our Style' was brought into the House of Lords by the Earl of Chesterfield. It had been recommended by Lord Macclesfield, the most eminent astronomical expert of the day, and proposed that the Julian calendar should be replaced by the Gregorian calendar.

The substance of the Bill was that January 1 should be generally accepted as the start of the New Year and that in 1752 'the natural day next immediately following the said second day of September shall be called reckoned and accounted to be the fourteenth day of September omitting, for that time only, the eleven intermediate nominal days of the common calendar'. It was also proposed that England should follow the Gregorian style and exclude three out of four end-century years from the leap year rule. The Bill, anticipating possible objections, adds, 'Nothing is intended to extend, to accelerate or anticipate the time of payment of any rent, annuity, or sum of money . . . but that all such rents, annuities, and sums of money . . . shall commence, cease, and determine at and upon the several days and times as the same should and ought to have been payable or made, or would have happened in case this Act had not been made.' The Bill was carried on 18 March 1751.

The *Gentleman's Magazine* cheerfully supported the Bill and in the year that it became law, 1752, celebrated the reform in fulsome verse exclaiming:

Man's superior Part
Uncheck'd may rise, and climb from Art to Art:
Trace Science then with Modesty thy Guide:
Instruct the Planets in what course to run,
Correct old Time and regulate the Sun.

In more extravagant lines the editor, whose pen-name was Sylvanus Urban, proudly pointed out that he had long ago advocated changes in the calendar and, though his proposals were at first

Lord Chesterfield who brought the Bill for reforming the calendar into the House of Lords in 1751. From an engraving in Doctor Johnson's House in Gough Square. (Dr Johnson's House Trust)

ignored, he was delighted that they had now at last been accepted:

Meeting Time the other day,
(He [the editor] knew that Time had lost his way)
Friend Time, says he, will not that Stile
Increase your journey half a mile?
Suppose this new rais'd path you try?
Time kept his way nor deign'd reply.

Time, the editor felt, would show his gratitude to the *Gentleman's Magazine* for its encouragement to change the style by preserving it together with the writings of Sylvanus Urban long after other magazines had died.

The whole of England did not accept the calendar changes so gleefully. The Duke of Newcastle begged Chesterfield before he introduced the Bill not to 'stir matters that have long been quiet' nor to 'meddle with new-fangled things'.

A ballad against a later Bill to do with the naturalisation of Jews recalled hostility to the Gregorian calendar because it had been devised by the Pope and the Catholics:

In seventeen hundred and fifty three
The style it was changed to Popery.

The religious were troubled because they felt that it was sacrilegious to change the saints' days. Others believed that their lives had been shortened by Act of Parliament which they thought had stolen from them rents and wages as well as time. Mobs rioted in London shouting, 'Give us back our eleven days.' They were bewildered that the Government could apparently get rid of eleven days.

The *Gentleman's Magazine* printed a letter saying: 'I went to bed last night, it was Wednesday September 2, and the first thing I cast my eye upon this morning at the top of your paper was Tuesday September 14.' Can he, the writer asks, have slept away eleven days in seven hours? Can the British Parliament annihilate Time? He quotes a line of verse, 'Ye Gods annihilate but Time and Space, and

'The Election', engraved from the painting by Hogarth. The man in the foreground having drink poured on to his head has his foot on a poster which says, 'Give us our eleven days'. (The Mansell Collection)

make two lovers happy,' which once seemed a wildly extravagant idea. Now: 'The British Parliament without the help of gods, or devils either, has accomplished one half of the miracle, and I'm convinced it will be found as easy to bring about the other.' The writer is bothered about celebrating anniversaries. When should he observe the feast of St Enurchaus which falls on September 7: 'The very next day to it is the nativity of Mary . . . I suppose she was to be born no day at all this year.' The writer himself has a birthday on September 13. When will he celebrate that? It is a problem familiar even today to those people, like James, the son of Princess Alexandra, who are born on February 29 and whose birthdays come only when there is a leap year.

The *Gentleman's Magazine* tried to put people's fear of losing rent at rest by a scheme 'to avoid confusion by two quarter days'. Normally one of the quarter days fell on September 29 and it was then that many rents were due and many servants received their wages for the past three months (a quarter of the year). 'The nominal quarter day will still be September 29, but as September 29 will be anticipated eleven days the real quarter day will be October 10,' so, 'it is proposed that all rent and servants' wages be paid on September 29 new-stile, the nominal quarter day, deducting such part of the sum as is in exact proportion to the eleven days that shall be wanting to complete the quarter.' With this advice the magazine printed a table to show how much should be taken from servants' wages according to how much they earned.

Resentment against the change lasted long after the reformed calendar had taken effect. In 1754, when Lord Macclesfield's son stood as parliamentary candidate in an Oxfordshire by-election crowds clamoured at his meetings for the return of the eleven days his father had stolen from them.

After 1752 only Russia of all European countries retained the Julian Calendar and kept it unreformed until the Revolution of 1918. In 1700 Peter the Great began to date the year from the birth of Christ and to start each year on January 1; before 1700, years had been dated 'from the Creation of the World' and New Year's Day had been on September 1. But these were minor changes. When, in 1918, the Russians adopted the Gregorian Style they had to drop thirteen days from the year to bring it into step. This has the odd result that the October Revolution is now celebrated in November. The Russian Orthodox Church continues to observe the Julian Calendar and so celebrates Christmas Day on January 7 (new style). Russia is not alone in its confusion. George Washington was born on 11

February 1732 (old style) but the Americans keep his birthday on February 22 (new style).

The most complete calendar change attempted since the Gregorian Calendar was that made by the French after the Revolution. The revolutionaries wanted to rid France of Christianity as well as of kings. They decided to date their years from 22 September 1792, the day when they abolished the monarchy, and also, by chance, about the time of the autumnal equinox. A man calling himself Marechal the Atheist had, ten years before, advocated a new calendar 'free at least of superstition' (Christianity was superstition to him), but Gilbert Romme was the principal creator of the French Revolutionary Calendar. A year was to be made up of twelve months of thirty days each. Sundays and all Christian Festivals were abolished. The weeks were ten days long with each day known simply by its number. The tenth day, the decadi, was a day of rest. To make up a year of 365 days there were five intercalated ones called Sans Culottides in honour of the Sans Culottes or working class extremists who had helped to bring about the Revolution. These five intercalated days were rest days and were held as Festivals of Genius, Labour, Actions, Rewards and Opinion. Leap Year day became a Festival of the Revolution.

The months were simply numbered first, second, third, until a poet, called poetically Fabre d'Eglantine, devised names for them descriptive of their characteristics. Thomas Carlyle in 'The French Revolution' lists the new months with mocking translations. The year began in our September with 'Vendémiaire, Brumaire, Frimaire, or, as one might say in mixed English, Vintagearious, Fogarious, Frostarious, these are our three Autumn months. Nivose, Pluviose, Ventose, or say, Snowous, Rainous, Windous, make our Winter season. Germinal, Floréal, Prairial or Buddal, Floweral, Meadowal, are our Spring seasons. Messidor, Thermidor, Fructidor, that is to say (dor being Greek for gift) Reapidor, Heatidor, Fruitidor are Republican Summer.'

Officially this calendar was used until 1806; but few people outside the administration kept to it. People could not easily alter the habit of centuries and the new calendar was unpopular because it changed the times of fairs and days of payment. Its use had to be enforced by the threat of prosecution. In 1805 Napoleon asked Talleyrand to convey to a colleague 'my displeasure at his dating his letter August 12 and at his not using the French calendar; while a law exists it must be obeyed . . .' The law did not exist for much longer.

The Labours of the months. A set of blue, white and yellow enamelled roundels designed by the Florentine Luca della Robbia to decorate a ceiling. 1345–50. The border in light and dark blue

shows the changing lengths of the days and nights from month to month. Behind the sun is the sign for the part of the Zodiac it is in during each month. (Victoria & Albert Museum, Crown copyright)

Napoleon abolished the Revolutionary Calendar to please the Pope and to help restore France to Roman Catholicism. From 1 January 1806, France started again to observe the Gregorian Calendar.

From time to time throughout the twentieth century there have been active movements to reform the calendar again. In 1930 Elisabeth Achelis devised her World Calendar and began working hard to get it accepted by all countries. She objected to the erratic month lengths of the Gregorian Calendar; most of us have to recite 'thirty days hath September' before we can be sure how many days there are to a certain month. In her book on calendar reform Elisabeth Achelis gives two examples of mistakes which can still easily slip through. In 1958 an advertisement for SABENA, the Belgian Airline, which was issued in five languages, gave September thirty-one days. And the Yale University Calendar for 1959 gave February thirty-one days. She wants the days of the month always to fall on the same day of the week; if you were born on Wednesday, January 4, your birthday would always be on a Wednesday. She feels that the study of history could be greatly enriched if we always knew without difficult calculations the day as well as the date of great battles or the death of kings. Finally with a more regular calendar there would no longer be great complications when we are 'making analyses, recordings, compiling data and planning engagements'. The numbers of hours worked in a month can vary considerably since some months sometimes have four weekends and sometimes have five. This can make things awkward for large firms employing a lot of workers.

The World Calendar smoothes out all irregularities. January 1 always falls on a Sunday. The year is divided exactly into four three-month periods, or quarters, of ninety-one days each. January, April, July and October, the first months of each quarter have thirty-one days each; the other months are thirty days long. This makes a total of 364 days. The

Elisabeth Achelis's World Calendar. The world's days between December and January and, in Leap Year, between June and July are clearly shown to be outside the regular run of days.

January					
Sun	1	8	15	22	29
Mon	2	9	16	23	30
Tue	3	10	17	24	31
Wed	4	11	18	25	·
Thu	5	12	1	26	·
Fri	6	13	20	27	·
Sat	7	14	21	28	·

April					
Sun	1	8	15	22	29
Mon	2	9	16	23	30
Tue	3	10	17	24	31
Wed	4	11	18	25	·
Thu	5	12	19	26	·
Fri	6	13	20	27	·
Sat	7	14	21	28	·

February					
Sun	·	5	12	19	26
Mon	·	6	13	20	27
Tue	·	7	14	21	28
Wed	1	8	15	22	29
Thu	2	9	16	23	30
Fri	3	10	17	24	·
Sat	4	11	18	25	·

May					
Sun	·	5	12	19	26
Mon	·	6	13	20	27
Tue	·	7	14	21	28
Wed	1	8	15	22	29
Thu	2	9	16	23	30
Fri	3	10	17	24	·
Sat	4	11	18	25	·

March					
Sun	·	3	10	17	24
Mon	·	4	11	18	25
Tue	·	5	12	19	26
Wed	·	6	13	20	27
Thu	·	7	14	21	28
Fri	1	8	15	22	29
Sat	2	9	16	23	30

June					
Sun	·	3	10	17	24
Mon	·	4	11	18	25
Tue	·	5	12	19	26
Wed	·	6	13	20	27
Thu	·	7	14	21	28
Fri	1	8	15	22	29
Sat	2	9	16	23	30

odd remaining day comes after Saturday, December 30. It is out of the scheme of days and months. It is simply a spare day not called a day of the week but named Worldsday or December W—to be kept by the whole world as a holiday. In leap year there is another holiday like this after June 30 called June W. Christmas Day always comes on a Monday.

In 1914 the Swiss Government was asked to investigate calendar reform on these lines, but the First World War put an end to the work. In the 1930s the League of Nations, after a week's conference on the matter, referred it back to the governments of the member countries for study and consideration. War in 1939 again put an end to any

July					
Sun	1	8	15	22	29
Mon	2	9	16	23	30
Tue	3	10	17	24	31
Wed	4	11	18	25	·
Thu	5	12	19	26	·
Fri	6	13	20	27	·
Sat	7	14	21	28	·
October					
Sun	1	8	15	22	29
Mon	2	9	16	23	30
Tue	3	10	17	24	31
Wed	4	11	18	25	·
Thu	5	12	19	26	·
Fri	6	13	20	27	·
Sat	7	14	21	28	·

August					
Sun	·	5	12	19	26
Mon	·	6	13	20	27
Tue	·	7	14	21	28
Wed	1	8	15	22	29
Thu	2	9	16	23	30
Fri	3	10	17	24	·
Sat	4	11	18	25	·
November					
Sun	·	5	12	19	26
Mon	·	6	13	20	27
Tue	·	7	14	21	28
Wed	1	8	15	22	29
Thu	2	9	16	23	30
Fri	3	10	17	24	·
Sat	4	11	18	25	·

September					
Sun	·	3	10	17	24
Mon	·	4	11	18	25
Tue	·	5	12	19	26
Wed	·	6	13	20	27
Thu	·	7	14	21	28
Fri	1	8	15	22	29
Sat	2	9	16	23	30
December					
Sun	·	3	10	17	24
Mon	·	4	11	18	25
Tue	·	5	12	19	26
Wed	·	6	13	20	27
Thu	·	7	14	21	28
Fri	1	8	15	22	29
Sat	2	9	16	23	30

W

chance of reform. In 1956 the World Calendar was presented to the United Nations as suitable for adoption by the whole world. But the committee considering it did not approve it and so it was not discussed in the Assembly. Yet there is still a World Calendar Association working to have it adopted.

In 1968 Christmas Day fell on a Wednesday. Many people did not return to work on Friday, December 29, for just one day before the weekend; they took an extra day's holiday. This was unfortunate because economically 1968 was a bad year for Britain and everybody was supposed to work hard to increase exports. So in March 1969 Brian Parkyn, Labour M.P. for Bedford, put down a private member's motion in the House of Commons urging the adoption of the World Calendar, which would fix Christmas perpetually on a Monday.

But in these days of swift travel and increasing trade a new calendar would have to be adopted simultaneously by all countries. Some people, like Orthodox Jews, have religious objections to the blank day of the World Calendar. Most people simply oppose change; our present system may, in some ways, be inconvenient, they say, but we know it and we are used to it. We manage; why change?

Chapter 3   Sundials and Sand-glasses

English sand-glass measuring one hour. Made in about 1650. In Flamsteed House, Greenwich. (National Maritime Museum)

The gods confound the man who first found out
How to distinguish hours—confound him too
Who in this place set up a sundial
To cut and hack my days so wretchedly
Into small pieces.

This is the complaint of a Roman poet, Aulus Gellius, who lived in the second century A.D. A sundial had recently been set up in the centre of the town where he lived, and a trumpeter or crier announced the hour that it told. Aulus Gellius wants the old days back when he could eat as his 'belly told him to', whenever he felt hungry. He is indignant that he must wait for the sundial to tell him that it is an appointed meal-time. The Romans had adopted the sundial from the Greeks, who probably brought it from Egypt.

The first recorded sundial is in the Bible, 2 Kings XX verses 9–11, where there is a reference to the shadow of the sun falling 'on the dial of Ahaz'. Ahaz had been king of Judah in 742 B.C. and the dial which he set up was still in use during the reign of his son Hezekiah.

We do not know what this biblical sundial looked like, but since it is called a dial it was certainly different from some of the early Egyptian sun clocks, parts of which are still in existence. In the Berlin Museum there is a fragment of an Egyptian shadow clock of about 1500 B.C. and there is a whole specimen from about 900 B.C. This is simply a horizontal wooden bar with a short upright and cross piece at one end. At dawn the device was placed with the cross piece towards the east so that its shadow fell on the bar where a

The Science Museum's copy of an Egyptian shadow clock of the 10th to 8th century B.C. From the original in the Neues Museum, Berlin. The hour marks give names to the different hours from the 'hour of rising' at the far end to the 'hour of high standing' nearest to the cross bar.

Two Egyptian women comparing the time by a shadow clock and a clepsydra (see page 55). Picture from 'An Illustrated History of Science' by Dr E. Sherwood. (Science Museum)

scale of six hours was marked. After noon the bar had to be turned round to show the remaining six hours of the day.

The hour then was not of the unchanging length we know today. The Egyptians divided the day and the night into twelve equal parts. Except at the spring and autumn equinoxes (the only two occasions in the year when the night and the day are of equal length), a day hour and a night hour were of different lengths and these different lengths changed throughout the year because the amount of daylight and dark varies according to the season —though the variation is smaller in Egypt than it is, for example, in Great Britain.

The Egyptians made another very simple shadow clock by fixing a palm rod upright in a piece of open ground. Twelve stones encircled the rod and the position of the rod's shadow on the stones indicated the time of day.

From such primitive beginnings many different kinds of sundial were developed. Some were engraved on walls, some were set up on pedestals, and in the sixteenth century portable sundials became fairly common.

In Roman times the *hemi-cycle* was a favoured kind. Many examples still survive. The hemi-cycle is said to have been invented by Berosus, a Chal-

dean astronomer, in 300 B.C. A semi-circle was
hollowed out of a block of stone. From the centre
to the edge of the semi-circle radiated eleven hour
lines cut into the stone. A metal pointer stuck out
horizontally from the centre of the semi-circle and
its shadow told the time, in much the same way as
the Egyptian palm rod.

A sixteenth century historian, John Stow, wrote
that in 606 'dyed St Gregory; he commanded clocks
(bells) and dials to be set up in churches to distinguish the hours of the day'. Dials from Saxon
times can still be seen on the walls of some old
churches. There is a famous one on Kirkdale

*Below left: Plaster cast of a
Roman "hemicycle" sundial.
(Science Museum) Below: The
original hemicycle which was
found at Civita Lavinia. Now in
the British Museum. The gnomon
is lost but the hole at the top where
it was fixed is clearly visible.*

church, near Helmsley, Yorkshire, which can be dated between A.D. 1055 and A.D. 1064.

The Saxons divided the day into four *tides;* the word in this sense still survives in *noontide* and *eventide.* It is the four tides which Saxon dials are designed to indicate. On a stone slab, mounted usually on a south wall, a horizontal line was engraved for the sunrise and the sunset; a vertical line indicated midday; the spaces between sunrise and midday, midday and sunset were also divided by a line. At the place where these lines met was a hole into which the gnomon or indicator could be pushed so that its shadow fell on the face of the stone. Some of these sundials seemed intended to show when it was time for services in the church rather than the time of day.

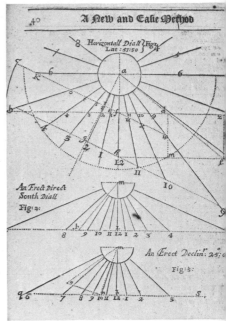

In the sixteenth and seventeenth centuries when portable dials made of ivory, brass, silver or bone became common throughout Europe, *dialling* was studied in schools and universities as a necessary part of a gentleman's education. In 1688 Thomas Strode, a mathematician, wrote a little book called 'The Art of Dialling' which taught 'how to measure the time of day by the shade of the sun'. William Molyneux, a Fellow of the Royal Society, at about the same time dedicated to Henry, Earl of Clarendon, his description of 'A new Contrivance adapting a Telescope to an Horizontal Dial for observing the moment of Time by day or night'. He had found the usual dials inconvenient because they had no division into minutes, half minutes and quarter minutes and 'everyone interested in Astronomy

Above left: Portable ivory tablet sundial by Paul Reinman of Nuremberg, 1599. The shadow cast by the string on the horizontal and vertical faces shows the time in the morning and afternoon. There is a compass for orienting the dial and four small dials which show the hour according to other systems of time reckoning. (Science Museum.) Above: Illustration from Thomas Strode's book 'A new and easy method of dialling', 1688. (National Maritime Museum, Greenwich) knows of what great Concernment the Observation of the exact moment of Time is Therein'. For without exact timing no exact observation could be made and astronomy could make no advances.

Molyneux points out that some dials do show minutes but then they have to be very large with a long gnomon which can cast only 'an uncertain shadow'. Also they can be used only on days when the 'sun shines out intensely'.

Though, as Molyneux says, minutes were important to early astronomers, they did not become really important to ordinary people until the Industrial Revolution quickened the tempo of life

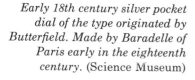

Early 18th century silver pocket dial of the type originated by Butterfield. Made by Baradelle of Paris early in the eighteenth century. (Science Museum)

with trains and factories. Now seconds and even fractions of seconds can be of vital importance to doctors and scientists concerned with complicated experiments.

Minutes and seconds are, like the hour, entirely artificial divisions of time. It was the Babylonians who first made use of the number sixty as a standard measure both in civil affairs and in astronomy. They regarded it as a mystic number, but it was also very useful because it can be divided by a larger number of figures than any other lower number. So from the Babylonians we get our sixty seconds in a minute and sixty minutes in an hour.

Molyneux claimed that his dial would tell the time by day and night. In fact since about 1520 nocturnals or night dials had been in use. In 1593, Thomas Fale wrote a book called again 'The Art of Dialling' where he devoted a chapter to 'the making of a dial, to know the houre of the moon'. Other nocturnals were made to tell the time by the position of certain moving stars in relation to a fixed one like the Pole Star. For just as the sun appears to turn around the earth giving us a solar day—that is a day measured by the sun—so the moving stars appear to turn around the Pole Star giving us a sidereal day—a day measured by the stars. [A sidereal day is four minutes shorter than a solar day.] So the *pointers* of the Great Bear constellation or the bright star of the Little Bear, for example, can be seen as the hour hand of an imaginary clock revolving around the Pole Star.

In the Science Museum in London there is a nocturnal of about 1700. It consists of a circular plate with a hole in its centre and a handle attached. Fixed to the plate is a smaller revolving one with two small projections marked G.B. or L.B. (so that

Wooden nocturnal, or night dial, of about 1700. (Science Museum)

you could find the time either by observing the Great Bear or the Little Bear), and a long revolving pointer extends far beyond the fixed plate.

To use the nocturnal you set the small revolving plate to the correct date using the projection marked G.B. or the one marked L.B. You then have to sight the Pole Star through the centre hole and turn the long pointer until its edge is in line with the star being observed. You can then read the time by the position of the pointer's edge on a scale of hours marked on the inner plate.

The Egyptians had observed the stars to tell the time at night but to measure the hours as they

Below: Late 17th century nocturnal in steel with elaborate decorative chasing. The months are not named, but are indicated by the appropriate sign of the zodiac. (Science Museum) Below left: How to use a nocturnal. Diagram from the Science Museum.

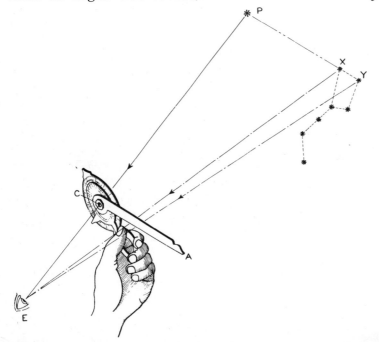

Cast of an Egyptian water clock or clepsydra from Karnak Temple. About 1400 B.C. The truncated cone shape ensures that the water level falls at a uniform rate.
(Science Museum)

passed they sometimes used *clepsydrae* or water clocks. One of the most simple kinds of water clock was a bowl with a scale of hours marked on the inside; there was a different scale for each month to allow for the changing length of the hour. The bowl had a small hole at the bottom. Through it water, which had been poured into the vessel to the highest hour mark at sunset, could very slowly trickle. By the falling level of the water men could tell the passing of time.

Pompey, the great Roman general, used a clepsydra, which he had captured in a war with the Eastern nations, to limit the speeches of Roman orators. When Julius Cæsar came to England he found a similar one which he used to observe that summer nights were shorter there than in Italy.

Some primitive Saxon water clocks found in the British Isles depend on a vessel filling rather than emptying. A bronze bowl with a small hole in its base was floated on some water. It gradually filled and sank, and the period of its filling and sinking was taken as a measure of time.

From these simple bowls the water clock developed into a most complicated piece of machinery. In A.D. 807, for example, the King of Persia, Haroun Al Rashid, presented a bronze water-clock, inlaid with gold, to Charlemagne, King of the Franks and

55

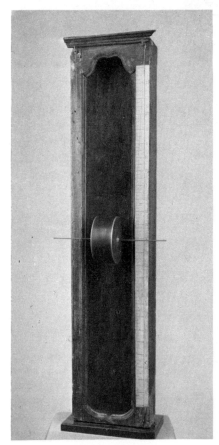

Emperor of the West. There was a dial with twelve small doors which represented the hours. 'Each door opened at the hour it was intended to represent and out of it came the same number of little balls, which fell one by one at equal intervals of time on a brass drum. It might be told by the eye what hour it was by the number of doors there were open and by the ear by the number of balls that fell.' Perhaps this was worked by a kind of water wheel like a famous Chinese water clock of A.D.1088. The inside of the wheel was divided into compartments. From outside a gentle flow of water was fed into the machine. This water dripped slowly from compartment to compartment, turning the wheel and the dial with it. In some later mechanical clocks it was the dial which revolved in this way and not, as now, the hands.

That same Charlemagne who was given the elaborate water-clock, ordered for himself a huge sablier or sand-glass. It had the hours marked on the outside and needed to be turned only once in twelve hours. It must have been huge indeed compared with the sand-glasses we use for timing eggs today. Sand-glasses were probably invented during the fourteenth century. The earliest illustration of one is on a wall painting in Siena, Italy, dated between 1337 and 1339. The subject of the painting is good government and one of its figures is Temper-

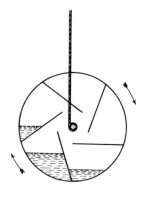

Left: English drum water clock of a type introduced about the middle of the 17th century. The diagram below shows how the weight of water moving from one section to another inside the drum turned it and made it move down the wooden case. (Science Museum)

ance holding a sand-glass. Besides being used in church to time sermons, sand-glasses were most frequently used on board ship to time the watches. The twenty-four hours of the day were divided into six watches of four hours each. So the ships' sand-glasses, which usually hung from ropes, were designed to run for four hours.

Much smaller, portable glasses usually ran for only four minutes, like a beautiful one in the Science Museum in London, which is mounted in silver. This is dated as late as 1799.

Besides sun, sand and water, people sometimes used fire to keep account of time. King Alfred, according to Asser, his biographer, had candles made which burnt for four hours so that he could know how long he had worked. And, in the thirteenth century, another king, Louis IX of France,

A fresco depicting good government by Lorenzetti in the Palazzo Publico in Siena shows the earliest illustration of a sand-glass; 1339. (Scala)

burnt candles of a fixed length so that he could time his reading.

Of these early timekeepers it is the sundial and the sand-glass which are most familiar. They continued to be used long after the invention of clocks. The sand-glass was especially useful on board ship because, unlike the early chronometers, it was not greatly affected by the pitching and rolling of the ship or by violent changes of weather.

When clocks and watches were newly invented they were extremely expensive, so the sundial remained popular; clocks were also inclined to be very inaccurate so the more reliable sundials were used to check them by.

It early became fashionable to engrave mottoes on sundials. There are sombre warnings that death is inevitable and terrible puns played on the word *dial* and the reminder that we *die all*. We are told, *Carpe Diem:* seize the present day. One sundial charmingly tells us that it measures 'golden hours only' while another boasts that:

A clock may mistake in the hour of the day
But the orb of the sun never goeth astray.

As late as 1804 some people still preferred sundials to clocks. The owner of a wooden one at the Redhouse Farm in Yorkshire said it was a 'good 'un to go by better nor any clock'.

Candle clock like the one King Alfred used, except that this measures twelve hours. (Science Museum.)

Chapter 4   Clocks

Of all the early time keepers it is the more elaborate clepsydræ, some of which had wheels driven by the changing weight of water, which are most closely related to early mechanical clocks.

The word *clock* comes from the late Latin *clocca* through French *cloche* meaning a bell. Bells were very important in the life of mediaeval townspeople. They rang the hours shown by sundials. They told people when to get up, when to go to bed, when to go to church. And they might warn of sudden disasters, like floods, or approaching invaders.

In London at the time of Shakespeare there were 114 churches. Most of these had bells, all with numerous excuses for ringing. Their noise could be most annoying to the sick and, in September 1592, 'extraordinary ringing' was forbidden. The writer of a book called 'Lachrymae Londoniensis' complained that 'in the night season (when we should take our rest) we are interrupted by the continual tolling of passing Bells, and anon the ringing out of the Sunne'. Passing bells were rung when somebody died. 'The ringing out of the Sunne' means that a bell was rung at day-break.

Possibly a striking mechanism was invented for sounding the bells and from this striking mechanism the large iron public clocks, which were the first true clocks, were developed. But exactly how clocks first came to be made is still uncertain and research

Many timekeepers of the mid 15th century. Behind the monk is a weight-driven mechanical clock and on the other side of the picture there is an automatic bell-striking mechanism. On the table are various portable sundials and the earliest known example of a table clock driven by a fusée (see page 72). From the manuscript of 'L'horloge de Sapience'. (Bibliotheque Royale de Belgique)

is difficult because a general word 'horologium' is used indiscriminately in early writings to refer to sundials, hour-glasses, water clocks, and any other kind of time keeper.

The first reliably recorded clock was set up in the church of St Gothard in Milan in 1335. It is described by an Italian chronicler as 'a wonderful clock with a very large clapper which strikes a bell twenty-four times according to the twenty-four hours of the day and night'. Other clocks which also struck the twenty-four hours were set up in Padua in 1344 and in Genoa in 1353.

Italy, with its system of counting the hours from one to twenty-four starting at sunset, differed at this time from the rest of Europe which, by the beginning of the fifteenth century, had adopted the two twelve-hour system counted from noon and from midnight.

The system of dividing the day and night into twenty-four parts comes from the Egyptians. At first they divided the night into twelve parts, each marked by the appearance of a particular star or constellation on the Eastern horizon. The hours of the day were numbered from one to ten according to the position of the sun; the twilight hours, one at dawn and another at dusk, were at first reckoned separately. Eventually things were simplified a little when the two twilight hours were counted in with the daylight hours and so the two twelve-hour system came into use. But the length of the hour still varied according to the time of year. Daylight lasts longer in the summer than in winter but as both the short winter day and the long summer day were divided into twelve equal parts the summer parts or hours were longer than the winter hours.

The Babylonians may have used a system with hours of equal length, but more certainly we know that at the beginning of the tenth century a Moslem astronomer made an unusual sundial that measured equal hours. With the invention of the mechanical clock equal hours came into general use. And it was from about 1500 onwards, when clocks were already beginning to be used but were too expensive for everybody, that people in Europe began to be interested in making sundials that could show equal hours.

In 1370 Charles V of France had a striking clock installed in a tower of the royal palace. He was so pleased with it that he had two more set up in other parts of Paris and he ordered all the churches to ring their bells when his clocks struck, so that everybody could know the correct time 'luise le soleil ou non' (whether the sun shines or not). The people of Paris were apparently less impressed than their king by this innovation. The clock in the palace was famous for its irregularity which inspired this little rhyme:

C'est l'horloge du palais

Elle va comme il lui plaît.

(The palace clock goes just as it likes).

One of Charles's clocks, made by the German clockmaker Henry de Vic, can still be seen on the side of the Palais de Justice in Paris. There are other very early clocks still in existence. One is in Salisbury Cathedral, one is at Rouen, in France, and another from Wells Cathedral is now in the Science Museum in London. These early clocks, massive, with their iron mechanism uncased and with only one hand to tell the hour, are hardly recognisable to us as clocks. The one at Rouen, set up in 1389, is the earliest clock known to strike the quarters as well as the hours, for though the one from Wells also has a quarter striking mechanism it was not set up until 1392. The clock at Salisbury which dates from 1386 is the oldest of the three. Although it no longer has a dial, it has been completely overhauled and restored and is still in good working order. The old elaborate dial of the Wells clock, which is probably not original, has been left at Wells Cathedral. On it are shown not just the hours of the day, but also the age and phases of the moon. When the hours strike, a number of figures on horseback perform a little show.

Such elaborateness was not unusual. There is another fourteenth century clock in Lund Cathedral, in Sweden, which tells 'among other things, the time, the date, the course of the sun and the moon'. At the stroke of noon and at three in the afternoon 'knights on horseback clash, trumpeters blow a fanfare, the organ plays "in dulci jubilo" and tiny doors open to admit small mechanical replicas of The Three Holy Kings as they file forward to pay obeisance to the Virgin mother and Child'. At Strasbourg and Bologna there are equally elaborate clocks.

Sometimes clocks which started out quite simply had more complicated additions made later. It seemed often to have been a matter of pride for a town to have an unusual clock. In Wimborne Minster, in Dorset, there is an astronomical clock which was made, it is claimed, early in the fourteenth century, when men still believed that the earth was the centre of the universe. So in this clock, as in others of its type found elsewhere, the earth is fixed at the centre. The moon moves round the earth and also turns on its own axis once every lunar month showing the phases by its colours of black and gold. The sun makes a larger circle around the earth and shows the time on a twenty-

The Lund Cathedral Clock. At noon the clock strikes and all the figures begin to move. Made in the 14th century. (Swedish National Travel Association)

The monk turned soldier who strikes the bells of the clock at Wimborne Minster in Dorset. 1613. (Photo: Royal Studios)

four-hour dial. Twelve noon is shown by a cross at the top and twelve midnight by another at the bottom. This dial is inside the church. On the outside of the tower a *jack* strikes the bell every quarter of an hour. A jack is a puppet-like figure. The one at Wimborne was added in 1613. 'To one of Blandford for carving of the Jack, 10s' says the churchwarden's accounts for that year. The jack started life as a monk but during the Napoleonic wars he became a soldier dressed in a bright red coat. More elaborately, in 1510, the figures of Adam, Eve and a serpent were added to the belfry of a clock in Ghent. Adam struck the hour, Eve struck the half hour, and the snake moved around the pair gloating over them.

It seems strange that all this effort was not put into making the clocks accurate rather than into the invention of ever more ingenious puppet shows. Would not a truly reliable clock have been a wonder in itself without elaborate additions? Now it is more important to have an accurate clock than to have one which tells the time roughly, together with the month, the day of the month, the number of days since the beginning of the year, the phases of the moon and the time of High Water at London Bridge. Henry VIII's astronomical clock at Hampton Court gives all this information. But in those days minutes were not felt to be very important.

Henry VIII never had a train to catch at 5.23 precisely.

Taxes were sometimes levied on townspeople for the purchase of a clock. In 1356 all the citizens of Bologna aged twenty and over were taxed eighteen pence each for a clock. And in Dundee in 1554 the Council and Deacon of Crafts ordered that 'an tax of 200 pounds be set and gathered for payment of the Knok (clock), and bringing home one bell for the same'. When the people of one town saw that their neighbours had a fine clock they were often eager to copy it.

Two more very early English clocks are one from Cassiobury Park, which is now in the British Museum, and one from Dover Castle, now in the

Science Museum. They are slightly smaller than those at Wells and Salisbury so it seems that they were made rather later. In construction they are exactly the same. The Cassiobury Park clock probably had to be wound about every six or twelve hours. Like some other early clocks it has no hands; instead the dial itself revolves.

The first domestic clocks, driven by weights and made all of iron, are simply smaller versions of such

Sir Thomas More with his family and his clock. Painted by an unknown artist in 1593. (The National Portrait Gallery)

public clocks. Again they were very expensive and they remained uncommon for many years. Kings and princes bought them, but more as rich men's playthings than serious time-keepers.

These early domestic clocks were generally called lantern clocks. Some people have suggested that this is because they are lantern-shaped, others say that lantern is a corrupt form of *latten* which was an old name for brass. At first domestic clocks were made of iron but they did come to be made of brass later. They could not stand on a table or cupboard because the weights had to hang free below. So either they were fixed to a wall or they stood on brackets which had two holes through which the cords that hold the weights could pass. The back and sides of the earliest clocks were open. There was generally only one hand and, as the clocks ran for only thirty hours, they had to be frequently rewound by pulling the weights back up with the opposite end of their cord. The hours struck on a gong at the top. There was rarely more than one clock in a house so it had to stand in a central position from which the strike could be heard in every room.

In the National Portrait Gallery there is a picture of Sir Thomas More, his family and descendants, painted by an unknown artist in 1593.

On the wall in the very centre of the picture is a lantern clock. Above the hour-dial there appears to be a dial showing the phases of the moon as well as the hours of the day.

Sometimes lantern clocks had an alarum which could be set on a small dial of Arabic figures. The alarum dial was usually inside the hour dial whose numbers were generally in Roman numerals.

But accurate time-telling was still not possible because of the crudity of the parts and because of friction. The King Charles who set up three public clocks in Paris during the fourteenth century owned several domestic clocks. Yet still, like Alfred of England and Louis IX of France in earlier times, he used graduated candles to measure time more certainly.

It was in about 1550 that lantern clocks began to be made of brass instead of iron. The movements, with wheels all of brass, began to be entirely enclosed in brass cases which were often elaborately engraved with curling leafy patterns. Usually there were still no minutes marked on the dials, for there was still no minute hand, but the spaces between the hour numbers were divided to show the quarters, and the half hour was often indicated by a *Fleur-de-lis* or some similar device.

Often the dials of these early clocks were marked

Late Medieval domestic Iron Clock. It has a painted dial. (British Museum)

with two sets of twelve hours rather than with the one set which is usual today. So Iago in Shakespeare's 'Othello' says of Cassio: 'He'll watch the horologe a double set—if drink rock not his cradle.' Meaning that as long as Cassio had not been drinking he could stay awake and on guard all day and all night: two 'sets' of twelve hours each.

While clocks were driven by weights they could not easily be moved about. It was not until springs instead of weights were used as the driving force that portable clocks and, later, watches became possible. Probably springs were used in a few clocks as early as the first part of the fifteenth century. To begin with, spring driven clocks were even less accurate than those driven by weights because there is greater driving force in a spring when it is tightly wound than when it is running down. So the clocks would go fast when they were first wound and then gradually slow down.

English Lantern Alarm Clock, made about 1660 by Thomas Knifton at The Crosskeys, Lothbury, London. (Science Museum)

Yet, apparently, men were proud to own such clocks and some even had them painted in their portraits. It is recorded that in 1481 Louis XI of France bought a clock probably driven by a spring since he intended to 'carry (it) with him in all places where he shall go'. It is shown in one of his portraits. The original picture has disappeared but there is an engraving from it in a book called

'Chroniques scandaleuses' published in Paris in 1620. It shows Louis XI standing; at his right on a table is the clock. It is narrow but quite tall, hexagonal in shape with the usual rounded dome of the striking clock. The British Museum has a French one very like this in its collection, dated 1528. A French writer on clocks, E. Gélis, tells an amusing little story about Louis XI's clock. One day there was a great crowd of people at the court and the clock disappeared. The King was not greatly worried, though it was not immediately found, because he knew that as long as nobody was allowed to leave, the thief would soon be betrayed by the striking of the clock.

Peter Hele or Henlein of Nuremberg is sometimes credited with the invention of spring-driven clocks in about 1500. But such evidence as that of Louis' picture suggests that they were in use before his time. A contemporary says of Peter Henlein, 'He is making small horloges of iron fitted with a number of wheels which, wherever they may be borne, and without a weight show and strike forty hours whether they be carried in the bosom or in the pocket.' This description makes them sound more like watches than clocks. Clearly they were smaller than Louis' clock. Because of their shape Peter Henlein's watches were called 'Nuremberg eggs'

Louis XI with his highly-prized portable clock. From an engraving in 'Chroniques scandaleuses', 1620.
(British Museum)

71

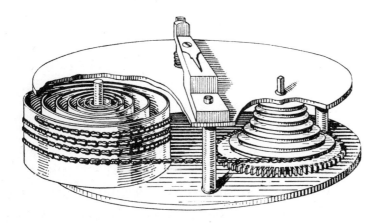

Drawing of fusée. (Science Museum)

and, though none of them has survived, there is written evidence that they were popular as royal presents made not just of iron but of gold or silver, often studded with precious stones.

A great step towards accuracy in spring driven clocks came when a man called Jacob Zech (Jacob the Czech) invented a device called the fusée in about 1525. Leonardo da Vinci had made drawings of a similar device thirty years before but we do not know that he ever used it in a time-piece. The fusée was a spirally grooved, tapering cone linked to the clock's spring, at first by a cord and later by a fine chain. As the spring uncoiled it pulled the cord, or chain, which thus turned the fusée and, through it, the clock's mechanism. 'When the spring is fully wound and exerting its greatest pull the chain is unwinding from the smaller end of the fusée, where it has only small leverage, while when the spring is nearly run down the chain pulls at the wider end of the fusée with a greater leverage.' This is how F.A.B. Ward in his book 'Time Measurement', published in 1958, explains the way the spring and fusée work together. The fusée thus solved the problem of a spring's greater force when it is tightly wound.

A clock made by Jacob Zech which employs the fusée now belongs to the Society of Antiquaries. It is drum shaped and has a horizontal face. It has two rings of hour numbers. On the outside ring Roman figures number the hours from I to XII twice; the inner circle numbers them from 1 to 24 in Arabic numerals. Around another circle are the signs of the Zodiac and a small indicator to show the phases of the moon.

Many early portable clocks had this drum shape or were like small square boxes. The horizontal face was often protected by an ornamented metal cover pierced with holes above the hour numbers to show the position of the pointer. Glass was not generally used to cover clock dials until about 1630.

Towards the end of the sixteenth century more portable clocks with vertical dials were being made. They were especially popular in France and Flanders. The earliest known English portable

Drum-shaped silver gilt table clock. Besides telling the time it indicates the position of the sun in the zodiac and gives the age, phase and aspect of the moon. German, probably from Nuremberg about 1590. (British Museum)

clock had a vertical dial. It was made by a man called Bartholomew Newsam in about 1580. He was appointed clockmaker to Queen Elizabeth I in 1583. His clock is square, rather solid looking, with richly engraved gilt, and it is on show in the British Museum.

In spite of these improvements clocks did not become really accurate until the invention of the pendulum and the hair spring.

It is to the Italian scientist Galileo that we owe the invention of the pendulum. He was born in 1564. While he was still a student he is said to have noticed the regularity with which a hanging lamp in Pisa Cathedral swung to and fro. He timed the swing by his pulse and found that it appeared to take the same length of time whether it moved in a small arc or a large one. From this observation he

Christian Huygens, the Dutch astronomer and mathematician, who first designed pendulum clocks. From an engraving in his complete works. (The Mansell Collection)

was able to design an instrument useful in medicine for timing pulse-beats. Towards the end of his life he drew designs for a pendulum clock. He never made a complete model because, with increasing age, he became blind.

It was not until 1657 that a pendulum clock was made by Salomon Coster of the Hague to the design of a famous Dutch astronomer and mathematician, Christian Huygens. Huygens designed his pendulum clock at first to be driven by weights but Coster later made spring-driven clocks with pendulums. Until the introduction of the pendulum, clocks had been regulated by a foliot. This is a horizontal bar with weights at each end which can be moved in or out to regulate the clock. When the clock is going, the foliot swings backwards and forwards. The pendulum made for greater accuracy because it is subject to the law of gravity—that is a natural force outside the clock itself—whereas the foliot and balance are governed entirely by variations in force within the clockwork.

The pendulum was found to be such an improvement that many clocks which had originally been made with the foliot and balance were later adapted to the pendulum. Both the Salisbury and Wells clocks have been adapted in this way; traces of the old foliot can still be seen. The Dover and

Cassiobury Park clocks are rare; they were not given pendulums but continued to be regulated by foliot and balance. In November 1692 Patrick Kilgour of Aberdeen was asked to 'translate the said (church) clock into one pendulum work conform to the newest fashion and invention done at London for regulating the motion of the said clock and causing her to go just'. While he was working on the clock he was also asked to make it 'strike the hours swifter that the people may not weary in telling of them'.

There are several domestic clocks in the British Museum which were also given pendulums when they came into fashion. On these the pendulum usually hangs down in front of the dial and because of its appearance is called a 'cow-tail pendulum'.

After various improvements had been made to Huygens' original design, chiefly to the way in which the pendulum was kept in motion, clocks began to be really reliable and capable of telling the time to the exact minute. So, in the seventeenth century, clocks with two hands instead of one began to be more frequently made. The first public clock in London to be made with two hands is that over the entrance door to St Dunstan's in the West in Fleet Street. Thomas Harrys made it in 1671 and in his specification he says, 'I will make two hands

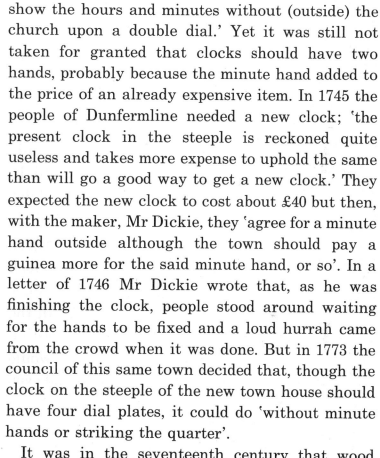

Left: Astrolabic table clock with a cow-tail pendulum. Made in Nuremberg about 1560. (British Museum) Below left: Thomas Harrys's clock outside the church of St Dunstan's in the West in Fleet Street. Made in 1671. The two giant jacks strike the hours and the quarters and turn their heads. (Shell-Mex and B.P. Ltd) Below: English long case clock by Henry Godfrey of London. Made about 1700. (Victoria and Albert Museum, Crown copyright)

show the hours and minutes without (outside) the church upon a double dial.' Yet it was still not taken for granted that clocks should have two hands, probably because the minute hand added to the price of an already expensive item. In 1745 the people of Dunfermline needed a new clock; 'the present clock in the steeple is reckoned quite useless and takes more expense to uphold the same than will go a good way to get a new clock.' They expected the new clock to cost about £40 but then, with the maker, Mr Dickie, they 'agree for a minute hand outside although the town should pay a guinea more for the said minute hand, or so'. In a letter of 1746 Mr Dickie wrote that, as he was finishing the clock, people stood around waiting for the hands to be fixed and a loud hurrah came from the crowd when it was done. But in 1773 the council of this same town decided that, though the clock on the steeple of the new town house should have four dial plates, it could do 'without minute hands or striking the quarter'.

It was in the seventeenth century that wood began to be used to encase clocks. Long-case clocks, or grandfather clocks as they are now commonly called, developed from lantern clocks. The long cupboard-like box enclosed the weights and pendulum keeping them from dust and the fingers of children. The case also made the clock less like a piece of machinery and more like a piece of furniture.

At first the case, made usually of plain oak, was simple and could most aptly be described as cupboard-like. But gradually the cases became highly decorated and ornamented according to the taste of the day.

The earliest grandfather clocks went for only thirty hours at a time, so that they had to be frequently wound. They gradually became eight day clocks, which went for eight days without winding. Presumably the length of eight days was chosen so that the head of the house could wind his clock every week, say every Sunday evening, and since it still had a day to go it would never stop. In the eighteenth century there was some competition among clockmakers to produce clocks which would run for a month, three months, six months, and even a year. The famous clockmaker Daniel Quare made a clock for William III which went for a year without winding. But it cost £1,500—about £30,000 today.

So that people could know the time at night when it was too dark to see the dial, a repeating mechanism was developed by a man called Barlow in 1676. The famous clockmaker Thomas Tompion made

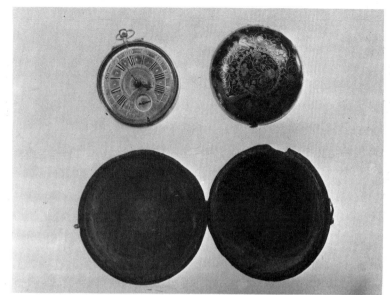

Pocket watch signed 'Thos. Tompion, London, 4'. Made c. 1685. The dial and case are of silver. The outer protecting case is of brass covered with horn inlaid with silver. Finally there is a contemporary leather case, but it was probably not made especially for this watch. (British Museum)

clocks which employed Barlow's invention. Usually there was a small string coming out of the case; when the string was pulled the clock would repeat the last hour or quarter it had struck. At first this mechanism was used only in clocks but later it was employed in watches too. Watches with a repeating mechanism came to be called simply *repeaters*. A clockmaker called Francis Torado used a simpler way of telling the time at night. The Museum of the Clockmakers' Company in London has an oval watch made by him between 1650 and 1660. It has a silver dial with an hour circle and one hand. Beside each number there is a raised pin so that at night the time can be discovered by touch.

Huygens, who had brought accuracy to larger clocks with the pendulum, also brought a great

79

improvement to portable clocks and watches with the introduction of the hair, or balance, spring. This fine spring coils and uncoils with a movement as constant as that of the pendulum. Its precision and compactness made pocket watches possible, so that in 1676, for example, a clockmaker called Daniel Le Count was able to make a watch only five and a half centimetres in diameter but accurate enough to tell the minutes as well as the hours.

Samuel Pepys often writes about watches in his diary; to him they were clearly new and of great interest. In an entry for the year 1665, for example, he writes, 'To my Lord Bruncker's and there spent the evening by my desire in seeing his Lordship open to pieces and make up again his watch, thereby being taught what I never knew before.' There seems to have been some exchange of watches between Pepys and Lord Bruncker, for in September 1666 Pepys writes, 'He (Lord B.) do now give me a watch, a plain one, in the roome of my former watch with many motions which I did give him. If it goes well I care not for the difference in worth though I believe there is above £5.' The words 'many motions' suggest, perhaps, that like Zech's small clock Pepy's 'watch' had told more than just the time of day. In Pepy's days the word *watch* meant more than it does now; it was often used of quite large time-pieces and the word clock was kept for those which struck or chimed. In a later entry Pepys writes of a 'larum watch', the equivalent of our alarum clock, which will help him get up 'betimes'. Pepys was given a watch as a present, 'and a good and brave piece it is . . . worth £14'. Fourteen pounds in the seventeenth century is the equivalent of at least £500 today.

A bearded gentleman holding a watch. Painted by Tommaso Manzuoli in about 1558. The watch is German and typical of the period. (Science Museum)

Though Pepys in the seventeenth century was clearly excited about watches and seems satisfied with their accuracy, Dr Johnson, about a century later, did not share his enthusiasm. He said of dictionaries that 'like watches the worst is better than none, and the best cannot be expected to go quite true':

But today we do expect our clocks and watches to go quite true. Towards the end of the nineteenth century electric clocks began to be developed in which electric currents were used to replace weights or springs as a source of power. In 1921 a man called W. H. Shortt made a new kind of pendulum clock run by electricity. It made an error of only a few thousandths of a second each day and was used in the Royal Observatory at Greenwich until 1942. This, in its turn, was replaced by a quartz crystal clock which can keep time to one thousandth of a second a day. Those who desired even greater accuracy developed the atomic clock which was completed in 1959 and 'has an error in time-keeping of only one second in 1,000 years'.

Quartz crystal clock. (Science Museum.)

Chapter 5   **Clock Makers and Menders**

At first, when the demand for clocks was not very great, clock making was not a specialised trade. Blacksmiths, locksmiths, or gunmakers made clocks too. In country places as late as the eighteenth century the blacksmith sometimes made the clock for his village church and kept it in repair. In the Dorchester Museum there is the old Bere Regis church clock. It was probably made by a blacksmith called Lawrence Boyce who also, in about 1710, made one for St Mary's Church in his own village of Puddletown. The records for Pilton Church in Somerset note that in one year the smith was paid three pence for repairing the clock.

Sometimes friars or priests devoted themselves to the making and mending of clocks. The town

*Works of the old Bere Regis clock.
1719. In the Dorchester Museum.
Lawrence Boyce Piddletown. Fecit
1719 inscribed across the top.*
(Dorset County Museum)

records of Aberdeen for 1537 say that the town
'knok' was 'reformed and mended' by Friar Alexander Lindsay. But some years later it seems that
Aberdeen did not even have a clock-mending friar
for, when it was found in 1618 that the three town
clocks were 'out of all frame and order and not
sufficient and able to serve the town', the magistrates were asked to 'write south' to find the best
'knokmaker' to come and mend them, or if necessary make new ones.

Usually these early public clocks had a specially
appointed 'governor' who had to keep them going
and, as was often necessary, re-set them according
to a sundial which, if the sun shone, could always
be relied upon. In the accounts of St Paul's Cathedral as early as 1286 there is an allowance made to

Sundial from St Paul's Cathedral. Used by Langley Bradley for correcting the clock. The sundial is 2 ft 5 inches in diameter and the fin is 19 inches high. (Dean and Chapter of St Pauls)

Bartholomew Orologario. From his name he was clearly the keeper of some kind of time-piece. His allowance was a portion of bread a day. Since the very earliest public clocks were made in Italy it has been suggested that the clock at the old St Paul's was imported from there together with the clock-keeper. It was without a dial but probably there were striking jacks. When Christopher Wren built the cathedral as we know it today, the maker of the clock, Langley Bradley, asked him to have a large sundial set up on the roof over the Library so that his clock could be checked by the sun. This sundial is now exhibited in the Library of St Paul's. In the Cathedral contract book for June 1675 to July 1709 it says that: 'Langley Bradley for himself and Executors etc. agrees to wind up (the clock) & look after, to repair & amend for 7 years from the 24th June next, or from the day of completion & to leave the same at the end of 7 years in good order and going condition.'

In the records of Aberdeen again, this time for 22 May 1453, it is written that 'there was granted . . . to John Crukshanks the service of keeping the orlage for this year and to have for his fee for the service of it XLs and has sworn the great oath to do his delligent business to the keeping of it'. In 1626 William Williamson was given the job of

85

'holding the knok in temper' and if the 'knok happened to be one quarter one hour out of temper' the unfortunate man was to lose a quarter of his year's pay. In 1680 John Mason of Bristol received one shilling a year to mend the clock of St John's Church; he undertook to keep it in order during his lifetime.

Carlo M. Cipollo in his book 'Clocks and Culture' gives us this account of a governor's work. 'Often the governor had to wind up the clock twice a day and had therefore to climb twice a day to the top of the clock tower; he had very frequently to grease the machine, because the gears were not so smoothly and precisely constructed; he had finally to re-set the hand (or the hands) of the clock almost every time this was being wound, because the clock lost or gained much time in the course of half a day.'

Winding the clock alone could be heavy work. A weight of one of the turret clocks made for Charles V of France is said to have been of at least 300 pounds. When in the 1890's Dean Gregory of St Paul's wanted Langley Bradley's clock replaced by a new one, his complaint was that Bradley's was so difficult to wind that it must have shortened the lives of several Cathedral workmen.

As the demand for clocks became greater, crafts-men began to specialize in making them. Paris was the first town to have a clockmakers' guild, in 1544. Craftsmen of all kinds used to form themselves into such guilds to protect and regulate their trade. They would set out terms of apprenticeship and decree a standard of workmanship. The apprenticeship for a clockmaker could take as long as seven years. Afterwards the young man had to spend two years as a journeyman in the workshop of a master clockmaker. As a journeyman he would

The interior of a clockmaker's shop. From an engraving by Philippus Gallaeus in 'Nova Reporta' published in Antwerp about 1600. (British Museum)

earn a wage but he was still not free to branch out on his own. After those two years he was expected to present a masterpiece; that was a clock of his own making which, if it reached the standard required by the guild, would qualify him to become a master clockmaker himself, able to have his own workshop, employ men, and train apprentices.

The Clockmakers' Guild of Paris recognised three different branches of clockmaking. The apprentice could choose one of them for his masterpiece. He could either make a pocket watch, or one of 'moyen volume', that is a medium sized clock, but still perhaps portable, or he might make a wall or tower clock with weights.

One apprentice's masterpiece was 'l'horloge à réveil-matin', that is an alarum clock, which also rang the hours and half-hours.

In London the Worshipful Company of Clockmakers of the City of London was set up on 22 August 1631, when King Charles I granted the clockmakers their charter. Their charter gave the Company control of all those 'either making, mending, buying, selling, ingrossing or retailing or (who) do any manner of work generally or particularly belonging to the Trade, Art or Mystery of Clockmaking'; it was their duty to prevent 'the making, buying, selling, transporting and importing any bad, deceitful and insufficient Clocks, Watches, 'Larums, Sun-dials, Boxes or Cases, for the said Trade, Art or Mystery of Clockmaking . . .' By the terms of the Charter the Master, Wardens and Assistants of the Company were empowered to 'enter into any ship bottom, vessel, or Lighters, or any other warehouse or warehouses, houses, shops or any other place or places whatsoever . . .' and they could search for all that are 'faulty and deceit-fully wrought, to seize on them and to break them, or cause them, if they shall think fit it may be well done, to be amended'. The power of the Company stretched within a ten mile radius of the City of London. It is especially interesting that the Company had jurisdiction over all timekeepers, not just over clocks and watches. They used their right to seize badly-made timepieces until the middle of the eighteenth century.

It seems that magistrates did not always accept the wardens' right to make these seizures. Those of Edinburgh fined and imprisoned Alexander Brownlie, a clockmaker, in 1721, after he had 'made seizure of a part of the dial work of ye town clock (this was a new clock that was being erected in St Giles Church, the manufacture of L. Bradley, London) as being wrought by an unfreeman'.

A book called 'Some Account of The Worshipful

Company of Clockmakers of the City of London' by two writers Atkins and Overall, published in 1881, describes some of the searches which were carried out in the cause of good clock-making. On 3 November 1652, a chamber clock which had been seized by the Warden, Mr Thomas Holland, was found 'very unserviceable and deceitful by the court' and therefore it was 'adjudged to be defaced and broken'. In 1671 two sundials were seized 'of Mr Frewen a seller of canes, etc, next the sign of the Naked Boy in Cheapside'.

Besides these searches the court could investigate the qualifications of those who set themselves up as clock- or watch-makers, as they did in the case of Peter Patmore in 1813. He had 'commenced trade to deal in, buy and sell watches and at the same time circulating printed Watch Papers and Shop Bills setting forth himself to the public as a Watchmaker', so he was summoned to appear at the monthly court. The man did not stand up to cross-examination, for 'on being questioned as to his practical abilities or skill in the Art of Clock or Watchmaking after some equivocation declared that he did not pretend to any such knowledge'. Yet in spite of his confessed lack of knowledge he became a master a few months later. Presumably he quickly made a masterpiece that satisfied the

Court and paid his admittance fee, 'to the use of the said Master, Warden and Fellowship, the sum of twenty shillings and to the clerk three shillings fourpence, and to the beadle twelvepence'.

In places outside London it was the City Fathers who made and administered the regulations to do with trading and the setting up of businesses. Strangers who came to the city had to prove themselves competent, but the rules about apprentices

Watch in gold with enamel dial. Mid 17th century. Signed on back plate 'Eduardus East, Londini'. (Victoria and Albert Museum)

were not so strict as they were in London. In October 1703 John Stretch petitioned the Common Council of Bristol to live in the city and work at clockmaking there. He was willing to buy the right either with cash or with a clock of his own making for the Council House. The clockmakers of Bristol did not welcome competition from him and they in their turn petitioned the Council against him. But the Council voted this 'as a very impudent and saucy petition' and eventually, in 1705, John Stretch was admitted a Free Burgess of the city after he had made a clock which he agreed to look after 'atte his own charge during his life'.

Atkins and Overall write also about some of the most celebrated freemen of the Clockmakers' Company. There was John Arnold who was born in Cornwall in 1744. He is especially famous for his chronometers and George III gave him £100 so that he could make experiments to improve them. In return Arnold made 'the smallest repeating watch ever attempted which he presented on June 4 as a birthday gift to George III. It was set in a ring and was smaller than a two-penny piece; it contained 120 different parts and weighed 5 dwts (penny-weights) $7\frac{3}{4}$ grains.'

Another celebrated master was Edward East who lived from about 1610 to 1693. It is said that Charles

II, when he was Prince of Wales, was fond of playing tennis. The prize for winning a match might be an 'Edwardus East' as the Prince called it, 'that is a watch of East's making'. Sir Thomas Herbert, Groom of the Chamber to Charles I, wrote in his memoirs that one morning he was late going to the King's bedchamber. The King said, 'You shall have a gold Alarm Watch which as there may be cause shall awake you; write to the Earl of Pembroke to send me such a one presently.' And, Sir Thomas says, 'The Earl immediately sent to Mr East.'

The Victoria and Albert Museum has an Edwardus East in gold and enamel shown on page 91. And the British Museum has an elegant table clock made by him in about 1665.

But clockmakers were not only interested in pleasing kings. Thomas Earnshaw, who was born in 1749, managed to make timekeepers that 'were so simple and cheap that they were within the reach of private individuals'.

The Fromanteels were a family of Dutch clock-makers who settled in England in the early part of the seventeenth century. John Fromanteel returned to Holland to learn how to make the pendulum clocks which had been invented by Huygens and made by Coster. When he came back he made, with other members of his family, the first pendulum clock in England. John Evelyn, the diarist, records that on 3 May 1661, he had visited 'Fromantil's ye famous clockmaker to see some pendules'. The British Museum has a long case clock made by Ahasuerus Fromanteel in about 1670.

Thomas Tompion, who has been called the 'father of English watchmaking', became a freeman of the Clockmakers' Company in 1671. He made barometers and sundials for William III and clocks for the Old Royal Observatory at Greenwich. One which he made for the Pump Room in Bath in 1709 is still in working order. This clock does not simply tell the time; it also gives the dates. Tompion was buried in Westminster Abbey when he died in 1713 and on his tombstone is a tribute to this maker of clocks 'whose accurate performance are ye standard of mechanic skill'.

In 1700 a kind of newspaper called 'The Affairs of the World' reported that 'Mr Tompion, the famous watchmaker in Fleet Street, is making a clock for St Paul's Cathedral which it is said will go for 100 years without winding up; will cost £3,000 or £4,000 and be far finer than the famous clock at Strasbourg'. It was just a wild rumour. Langley Bradley not Tompion made the clock for St Paul's. But it does show the wonders Tompion was thought capable of.

Thomas Tompion, 'the father of English watchmaking'. Portrait by Sir Godfrey Kneller, 1697. (British Horological Institute)

He is said to have left English watches and clocks 'the finest in the world'. By the end of the seventeenth century many English clockmakers were world famous for the accuracy and beauty of their timepieces. In 1704 a complaint was made to the Company of Clockmakers that some clockmakers in Amsterdam had put the names of Thomas Tompion and Daniel Quare on their timepieces so that they could cheatingly sell them as the works of these English masters.

The Worshipful Company of Clockmakers still thrives in London today. It no longer seeks out and

Mr Wilfrid Cripps winding the clock of St Paul's Cathedral in the days before it was wound electrically. (The Times)

destroys 'deceitful clocks' but many of its members continue to be active clockmakers.

Some public clocks still have 'governors' to wind them and look after them. But gradually the best known are being converted to automatic electric winding. Big Ben, which after many difficulties of construction started working in 1859, was hand-wound until 1913. This took a man thirty hours a week, which is not surprising since the clock tower is 316 feet tall and the weights fall almost to ground level when the clock is run down. Since 1913 an electric motor has done the winding. It now takes 1 hour, 45 minutes a week; 45 minutes every Monday and half an hour on each Wednesday and Friday.

Messrs E. Dent and Company, who made Big Ben in the nineteenth century, still maintain it and several other public clocks in London, a few of which continue to be hand-wound.

The clock at St Paul's Cathedral, not Langley Bradley's but its replacement made by J. Smith and Sons of Derby in 1895, used to take half an hour to wind every day. Mr Wilfrid Cripps was its 'governor' for thirty-four years. Daily he climbed the 200 spiral stairs to the top of the tower to wind up the 5 cwt quarter chimes weight and the 9 cwt hour weight. Then in July 1969 the clock was converted to automatic winding. The days of the governors will soon be over.

Time and Travel   Chapter 6

In 1675 Charles II founded the Royal Observatory at Greenwich 'in order to the finding out of the longitude of places for perfecting navigation and astronomy'. Before the discovery of the New World, navigators had chiefly kept to coastal routes and to the Mediterranean; now they were sailing vaster seas like the Atlantic and had to know their latitude and longitude if they were not to get hopelessly lost and possibly run aground. The longitude of a place is its angular distance east or west of a given point. Its latitude is the angular distance north or south of the equator. Navigators had long been able to work out the latitude by the position of certain stars in relation to the horizon. It was more difficult to find the longitude.

Above: John Flamsteed, the first Astronomer Royal. An engraving by George Vertue (1721) from a painting by Thomas Gibson. (National Maritime Museum) Below: Flamsteed at work with his assistant. Detail from the ceiling of the Painted Hall, the Royal Naval College, Greenwich. (National Maritime Museum)

John Flamsteed was the first Astronomer Royal and the first inhabitant of Charles' observatory. He took up his appointment in 1676 at the age of thirty, with the meagre salary of £100 a year and an allowance of £20 a year for an assistant. Flamsteed decided that an accurate Star Catalogue was the first step towards solving the problem of longitude. He worked incessantly to accumulate the enormous number of observations which were needed for true accuracy. He died in 1719 before his catalogue could be published, though against his will a pirated version had been brought out in 1712. For six years after his death his wife, Margaret, and two helpers worked together collecting his observations and in 1725 published them in three volumes called 'Historia Coelestis Britannica'.

This was his greatest work but he was also the first astronomer to make a methodical study of the difference between the time kept by clocks, which we call mean time, and that kept by the sun, which is called apparent solar time. The difference between the two is called the Equation of Time. There is a difference because the sun is not a completely regular timekeeper; the speed with which the earth turns in its orbit around the sun and on its own axis varies slightly throughout the year and so the apparent movement of the sun is irregular. We talk of the 'apparent' movement of the sun because it is of course the earth which moves around the sun not the sun around the earth. So clocks are made to work to the time of an imaginary sun which is regular in its apparent movement. The difference between mean time and apparent solar time varies from nothing to about six minutes fast or slow; there are just four times in the year when mean and solar time are the same. For his study of the Equation of Time, Flamsteed used two astronomical clocks, or regulators, made by Thomas Tompion. Replicas of them can be seen in the Octagon Room of Flamsteed House at Green-

Astronomers working in the Octagon Room of the Old Royal Observatory, Greenwich (now called Flamsteed House). From an engraving in the 'Historia Coelestis' by Flamsteed, published after his death. (National Maritime Museum)

LONDINUM.

The Old Royal Observatory from the south east as it looked in 1675. (National Maritime Museum)

wich. One of the originals is in the British Museum and the other is now privately owned.

A later Astronomer Royal, Nevil Maskelyne, sought an answer to the problem of longitude in the motion of the moon. In 1767 he brought out the first Nautical Almanac in which the Greenwich times when the moon would reach certain parts of the sky were predicted. Once the sailor knew the difference between local time and Greenwich time, which he could calculate with the help of the Nautical Almanac from the position of the moon, he could work out his longitude. In twenty-four hours the earth turns a full circle of 360 degrees on its axis. Therefore in one hour it turns fifteen degrees. If the ship was travelling west and there were three hours difference between Greenwich

The dial of Thomas Tompion's regulator which used to be on the wall of the Octagon Room, Greenwich. (British Museum)

time and local time the sailor knew that he had travelled forty-five degrees west of Greenwich and so he could find his position on a chart. After the first Nautical Almanac was published, map and chart makers began to use Greenwich as their starting point from which to measure longitude. So Greenwich began to be a place of great importance to many navigators.

Yet even with a Nautical Almanac navigators had to do many calculations before they could be sure of their longitude. What was really needed was a timekeeper which could be set to Greenwich time at the start of the voyage and which could keep to it accurately throughout. In the early eighteenth century a marine timekeeper of such accuracy hardly seemed possible. Sir Isaac Newton listed the difficulties: 'a Watch (was needed) to keep time exactly, but by reason of the Motion of the Ship, the variation of Hot and Cold, Wet and Dry, and the Difference of Gravity in different Latitudes, such a Watch hath not yet been made.'

Huygens in about 1662 had experimented with various kinds of marine timekeepers. But those with pendulums were not successful when the sea became rough because, to measure time accurately, a pendulum must swing from a fixed, unmoving support. They were upset too by changes in temperature and changes in the pull of gravity. Huygens also experimented with marine timekeepers regulated by a balance wheel and spring, but these were no more successful than those with pendulums.

Sailors had to endure many hardships, and some died because they could not work out their positions accurately. In 1707 there was a really spectacular disaster when four English ships and 2,000 men were lost including the Admiral, Sir Cloudsley Shovell. He mistook the fleet's longitude and ran into the Scilly Isles. Dismayed by this loss the British Government eventually decided to offer money prizes to anybody who could find a way of determining longitude accurately.

The idea of offering prizes of money was not new.

Sir Cloudsley Shovell. Illustration to an account of his life and poem lamenting his death published in 1707. (National Maritime Museum)

103

As early as 1598 Philip III of Spain offered a reward of 100,000 crowns to anyone who could solve the problem, and a little later the States General of Holland put up a prize of 10,000 florins. Some men felt that the problem could not be solved even by the devil; for a man to attempt it 'would undoubtedly be the height of folly'.

The British Government offered £20,000, worth about £350,000 today, to the inventor of a method which 'when tested by a voyage to the West Indies is found to be within 30 miles of the truth, £15,000 if within 40 miles of the truth, and £10,000 if within 60 miles of the truth'. A Board of Longitude was set up to test methods and to award the prizes. It was John Harrison, a Yorkshire carpenter, who finally managed to win the highest award of £20,000.

Harrison was born in 1693. He became a carpenter like his father, but he was interested in clocks. He had no formal training as a clockmaker but taught himself to repair them and then to make them. His knowledge of carpentry had its uses in clockmaking; many of his early clocks had all their wheels and pinions of wood.

In 1728 Harrison came to London with drawings of a marine timekeeper which he planned to make if, as he hoped, the Board of Longitude would give him some money. But before he took his drawings

John Harrison with marine timekeeper Number 4. From an oil painting by T. King. (Science Museum)

to the Board he visited the Astronomer Royal, Edmund Halley, at Greenwich. Halley was certain that the Board would not give him money on the strength of the drawings but sent him for advice to George Graham, the most important clockmaker in London at that time. Graham suggested that Harrison should make the timekeeper before going to the Board and even lent him money so that he could afford to do so.

John Harrison returned to Yorkshire where he worked on his marine timekeeper between 1729 and 1735. It was like a huge seventy-two pound, spring-driven clock controlled by two balances which were connected to each other by wires in such a way that they always moved in opposite directions. The movement of a ship could not upset these balances. There was also a device to compensate for changes in temperature.

Harrison went back to London with his timekeeper in 1736 and from there went on a trial run to Lisbon with it aboard H.M.S. *Centurion*. The Captain of that ship, Captain George Procter, wrote that Harrison was 'a very sober, a very industrious, and withal a very modest Man, so that my good Wishes can't but attend him; but the Difficulty of measuring Time, where so many unequal Shocks and Motions stand in Opposition to it, gives me

H.M.S. Centurion, the ship in which John Harrison went on a trial run with his first marine timekeeper. Model in the National Maritime Museum, Greenwich. (National Maritime Museum)

Harrison's first marine timekeeper at the back with Number 4 in the centre and others made by Larcum Kendal on the left and right. (National Maritime Museum)

concern for the honest Man, and makes me feel that he has attempted impossibilities'. Yet the trial of Harrison's marine timekeeper was so encouraging that the Board of Longitude allowed him a grant of £500 to help him improve it.

Harrison made two more very large timekeepers called Number 1 and Number 2 which, for various reasons, were not tested at sea. In 1757, while he was working on the third which he planned to enter for the £20,000 prize, he started on a much smaller marine timekeeper; it was only just five inches in diameter and he thought it might be a useful auxiliary to the much larger Number 3. When he finished this small watch he found it was just as accurate as Number 3 and had the advantage of being much more portable. So he entered the watch, called Number 4, for the great prize.

His new timekeeper was found to be astonishingly accurate on a trial voyage to Jamaica in 1761 and another to Barbados in 1764. Captain Cook took a duplicate on his second voyage in 1772, and it was only 7 minutes 45 seconds slow after more than three years at sea. Yet the Board of Longitude quibbled and delayed with different excuses the payment of the full prize. They made various advances and grants from time to time. They insisted that Harrison, though he was now very old and rather infirm, should make yet another timekeeper which would undergo more rigorous tests.

Diagram showing longitude.
(National Maritime Museum)

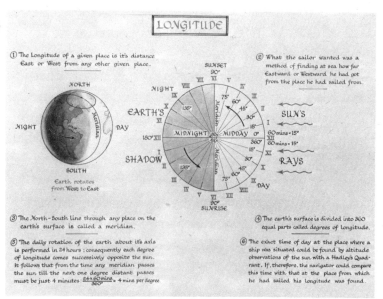

LONGITUDE

① The Longitude of a given place is it's distance East or West from any other given place.

② What the sailor wanted was a method of finding at sea how far Eastward or Westward he had got from the place he had sailed from.

③ The North-South line through any place on the earth's surface is called a meridian.

④ The earth's surface is divided into 360 equal parts called degrees of longitude.

⑤ The daily rotation of the earth about it's axis is performed in 24 hours; consequently each degree of longitude comes successively opposite the sun. It follows that from the time any meridian passes the sun till the next one degree distant passes must be just 4 minutes: $\frac{24 \times 60 \, mins}{360} = 4$ mins per degree.

⑥ The exact time of day at the place where a ship was situated could be found by altitude observations of the sun with a Hadley's Quadrant. If, therefore, the navigator could compare this time with that at the place from which he had sailed his longitude was found.

Harrison made another timekeeper with the help of his son. In its mechanism it was almost an exact copy of Number 4 but its appearance was less ornamental. It was tested by King George III in his private observatory at Richmond and in ten weeks it gained only four and a half seconds. Still the Board would not pay. When the King heard how Harrison had been treated he announced that in his opinion the man and his son had been 'cruelly wronged'. When Harrison finally petitioned Parliament for the payment of his money the Board found that both King and Parliament were against them. They could no longer avoid payment, so at last John Harrison was acknowledged the winner of the £20,000. In 1776, three years after he had won the prize, he died.

All Harrison's marine timekeepers, or chronometers as they are now called, and the duplicate (K.1), made by a London watchmaker Larcum Kendall and used by Captain Cook, are exhibited, going, in the Navigation Room of the National Maritime Museum. After years of neglect they were carefully restored and put into working order by Lieutenant-Commander Rupert T. Gould. The work took him from 1920 to 1933 and, like John Harrison before him, he came to it with no traditional clock-maker's training.

Chronometer of the 1840's made by E. Dent & Co. (National Maritime Museum)

John Harrison proved that accurate marine timekeepers were not 'impossibilities'. The success of this untrained Yorkshireman encouraged others to work at the problem for, though his Number 4 could hardly be improved for accuracy, it was a very complicated piece of machinery and consequently too expensive for general use.

It was a Frenchman, Pierre Le Roy, who, in 1766, quite independently of John Harrison produced a chronometer which is a true forerunner of all chronometers made today. And two English clockmakers John Arnold and Thomas Earnshaw are notable for their success in producing simple, cheap chronometers. Arnold who lived from 1736 to 1799 was at first unsuccessful. Three of his early timekeepers went with Captain Cook on his second voyage and they performed badly. But he persevered and finally was able to establish a factory at Chigwell which made accurate chronometers by the hundred. Unlike Arnold, Earnshaw did not massproduce timekeepers, but his chronometers were better; they were just as accurate, and simpler. They set the pattern for the modern chronometer. Lieutenant-Commander Gould says that if an Earnshaw chronometer of 1795 were put beside one made in the twentieth century it would be extremely difficult to find any difference in their appearance or in their mechanism. Thomas Earnshaw is commemorated in St Giles Church, Holborn, as 'the creator of the modern marine Chronometer'.

In 1821 the Royal Observatory began testing chronometers for the Royal Navy. We get some idea of these tests from 'The Horological Journal' for April 1893. They took place every year. They started on July 1 and lasted for twenty-nine weeks. The chronometers were put into an oven for two periods of four weeks each where the temperature ranged from 75 degrees to 100 degrees Fahrenheit. They were also tested at ordinary summer and winter room temperatures.

Greenwich, which had become important to sailors with the publication of the first Nautical Almanac in 1767 (it is still published today), became important to travellers on land also after the coming of the railways in the early nineteenth century.

Even after clocks became common there was no standard time for, as we have seen, clocks were put right by sundials which give local time and this varies from one place to another. For example high noon in Plymouth is sixteen minutes later than it is in London because Plymouth is west of London.

The people who ran and used stage coaches did

not concern themselves with minutes. Their time-
tables would seem exasperatingly vague to us;
a coach might be said to be leaving in the early
morning, about noon or towards sunset. Once
railways, with their swift travel, were introduced
this use of local time lead to great complications,
especially in a vast country like America. Each
railway company took its time from that of the
most important city on the line; in America there
were seventy-five different railways so there were
seventy-five different railway times. In addition
each community continued to keep its local time, so
there might be a great difference between the time a
railway company said a train left the station and
the time a traveller's watch, showing the local time,
said it left. In about 1880 the railway station at
Buffalo, in New York State, had three large clocks.
One told local Buffalo time, another New York City
time, for that was the time used by New York City
Railways, and the third told Columbus, Ohio, time
which was favoured by the Michigan Southern
Railway. And a traveller's watch might give yet
another time.

Clearly something had to be done about this
chaos. England decided to adopt a standard time.
For some years the makers of watches, clocks and
chronometers had been in the habit of visiting the

Observatory at Greenwich with watches so that they could check their regulators by the Greenwich clock. Then, in about 1836, John Belville, the adopted son of John Pond who was then the Astronomer Royal, was given the job of distributing Greenwich time. He began to call regularly on the chief London clock and chronometer makers with the correct time. When he died in 1856 his wife carried on. She retired in 1892 and her daughter, Ruth, began to make the weekly journeys to London distributing Greenwich time for a small fee. She came to be called the 'Greenwich Time Lady'. She always dressed in black and wore a large silver chronometer on a string attached to a safety pin. It had been made by John Arnold and she always called it Mr Arnold. She gave up the work in 1914 and from then until her death in 1943 at the age of 90 she received a pension from the Clockmaker's Company. It is surprising that she worked for so long since, in 1852, an electric master clock was set up in the Royal Observatory at Greenwich and a twenty-four-hour clock was mounted at the gate to the Observatory. The master clock controlled not only the gate clock but also the Greenwich Time Ball, a clock at London Bridge Station and a time ball at the offices of the Eastern Telegraph Company in the Strand. There had been a time ball

Ruth Belville's Mr Arnold.
(Guildhall Library)

at Greenwich since 1833 but at first it was not controlled by electricity. It was raised to the top of its fifteen-foot mast by a winch and at exactly one o'clock each day it was lowered as a time signal to sailors on the Thames and chronometer makers working in Clerkenwell. By 1852 the electric telegraph had been invented and it provided a way of distributing Greenwich Mean Time, or Railway Time as it was then called, throughout the country. The time balls were not always reliable even when they were controlled by electricity. 'The Horological Journal' for January 1892 says that the Greenwich Time Ball could not be raised on October 14, December 10 and December 13 of 1891 because of violent winds.

In 1893 the American railway companies held a General Time Convention. This convention divided the USA into four time zones which would each observe uniform time. America, of course, unlike England, is too big to keep one time throughout the whole country. The four zones are called Eastern, Central, Mountain and Pacific; they are based on Greenwich and there is an hour's difference between each of them. Eastern Standard Time is five hours behind Greenwich, Central six hours, Mountain seven hours, and Pacific eight. The dividing lines between the zones are not straight because they had to allow for State boundaries and railway terminals.

In 1884 there was an International Conference on Time in Washington. Countries had separately been attempting to standardize time; now it was felt that a world plan was needed. The conference decided that the Prime Meridian of the world should be at Greenwich; that world time should, as it were, start from Greenwich. This was an obvious choice since most of the world's ships already had the meridian of Greenwich on their charts and the Americans had based their time zone system on Greenwich Mean Time. Starting at Greenwich the world was divided into twenty-four standard time zones with an hour's difference between each. As travellers passed from one zone to the next they had to put their clocks and watches on one hour if they

The Prime Meridian, Greenwich.
(National Maritime Museum)

114

were travelling east, and put them back one hour if they were travelling west. A ship sailing from Southampton to New York crosses five time zones and therefore must put its clocks back one hour five times before it reaches its destination. When the time is 12 noon in Greenwich it is 7 a.m. in New York. Only two major countries did not agree to accept Greenwich as the prime meridian. Ireland kept Dublin time until 1916 and France perversely decided to keep a time 9 minutes, 21 seconds in advance of the meridian at the Paris Observatory. They thus avoided calling it Greenwich Time though it, in fact, works out to be the same as that of the meridian of Greenwich.

The Washington Conference also established the International Date Line which is halfway round the world from Greenwich. This is a necessary convention as ships travelling east around the world have added 12 hours and ships going west have lost 12 hours by the time they reach half way. So when they cross the dateline an adjustment of one day must be made to their calendars. When Mark Twain sailed west across the dateline he wrote in his diary:

'September 8. Sunday. Tomorrow we must drop out a day, lose a day out of our lives, a day never to be found again.

Next day. While we were crossing the 180 degree meridian it was Sunday in the stern of the ship where my family was and Tuesday in the bow where I was . . . Along about the moment we were crossing the Great Meridian a child was born in the steerage and now there is no way to tell which day it was born on.'

But if Mark Twain was confused, consider how much more confusing time has become in these days of swift air travel. A sea voyage is a fairly leisurely affair and gives the traveller a chance to adjust to changes in time. A man who flies from New York to Berlin gets six hours out of step with his daily life. His watch and body, set to New York time, may tell him it is 1.30 p.m., just after lunch, when he arrives but in Berlin he will have to accept that it is 7.30, coming up to dinner time and a whole afternoon lost. It is even worse for the traveller who flies from, say, Tokyo to San Francisco. If he leaves Japan at 2 p.m. on a Tuesday, his journey of almost ten and a half hours will take him across six time zones as well as the International Date Line. He will arrive in San Francisco at about 6.30 on Tuesday morning. So, if he feels up to it, he can eat Tuesday breakfast all over again as well as Tuesday lunch, though his body, which is adjusted to Tokyo time, tells him that it is the middle of the night and all he wants to do is sleep. It may well take him several days to get into step with his new time. J. B. Priestley writes in 'Man and Time' of his bewilderment when, on a flight across the central Pacific Ocean, 'in an unchanging clear morning light I was served what appeared to be a series of breakfasts'.

Though the coming of the railways helped to bring about standard time and made order out of chaos, it must sometimes now seem that the railways deliberately make things more confusing than they need to be. Whereas we normally divide the day into two periods of twelve hours each, the railways, like airlines and the armed forces, use a twenty-four-hour day; 2 p.m. becomes 14 hours and 9 p.m., 21 hours. This use of the twenty-four-hour clock avoids the need to put a.m. or p.m. after the time. The abbreviation a.m. stands for the latin words 'ante meridiem', meaning before noon; p.m. stands for 'post meridiem', meaning after noon.

With the Conference of 1884 it seemed that men had arranged time satisfactorily. But in 1908 William Willet, a Chelsea builder, had an idea which he thought would give people more chance to get out in the open air after work. He suggested that clocks should be put forward one hour in summer so that evenings would be light longer.

This was called his 'Daylight Saving' scheme. It was not used until 1916. By then England was fighting the First World War and it was thought that the daylight saving scheme would help to save fuel and would also give agricultural workers longer to work in the fields. A Bill was passed through Parliament stating that in the 'prescribed period in each year during which this Act is in force the time for general purposes in Great Britain shall be one hour in advance of Greenwich Mean Time'. Even after the war that extra light hour in summer was found useful, so in 1925 'Summer Time' became a permanent measure in Great Britain. Until 1939 the 'prescribed period' ran from 'two o'clock in the morning following the first Saturday in April . . . until two o'clock in the morning next following the third Saturday in October'. During and after the Second World War these dates varied and Summer Time was prescribed each year by an order in Council. During the 1940's Double Summer Time was introduced. This meant that the evenings were light even longer, for the clocks were put on two hours instead of only one.

On 18 February 1968, Britain put its clocks on for an hour and left them there. We were to have perpetual Summer Time with no return to Greenwich Mean Time. The Government had decided on

Flamsteed House as it is today showing the time ball and the gate clock which still shows Greenwich Mean Time. (National Maritime Museum)

a three years experiment. What used to be British Summer Time, which was in fact the same as Central European Time, should now be British Standard Time. It was felt that businessmen with interests in Central Europe would find life easier if our time was permanently the same as that of our neighbours, especially if Great Britain joined the Common Market. There would no longer be any need for hasty calculations to decide the time in Germany before a telephone call was made. Also it was thought better for children to be able to come home from school in the light during the winter as well as the summer. Unfortunately many children now have to go to school in the dark during the winter. In some parts of Scotland there is darkness until ten and even eleven in the mornings.

War changed Time and it changed the Observatory at Greenwich. In 1939 Greenwich's Time Department moved to Abinger in Surrey where it would be safe from bombs. After the war the entire observatory moved to Herstmonceux Castle in Sussex, where it is now called the Royal Greenwich Observatory as opposed to the Old Royal Observatory still at Greenwich. This permanent move became necessary because Greenwich, which in the eighteenth century had been a country village, had,

Herstmonceux Castle in Sussex,
now the home of the Royal
(National Monuments Record,
Crown copyright)

even before 1939, become crowded close by London with its smoke, fog and street lamps hampering observations. So now Greenwich Mean Time is controlled from Herstmonceux with allowances made for the slight difference in longitude between Greenwich and Herstmonceux.

Herstmonceux also now controls some useful services which were started at Greenwich. In February 1924 the 'six pips' time signal was introduced. This signal was sent out from a regulator clock at Greenwich through the BBC so that people had a reliable and accurate time check. These pips are still sent out on the various BBC sound programmes. In 1929, time signals controlled by Greenwich began to be sent by GPO radio to all parts of the world. And since 1936 signals by landline from the Royal Greenwich Observatory have corrected the telephone's 'speaking clock'. In Britain you can dial TIM or 123 on the telephone and a woman's voice tells you the exact time every ten seconds. The Post Office claims that it is normally correct to 0·05 of a second.

With all these changes it may seem that Greenwich has lost its important place in the world. But the Prime Meridian still passes through Greenwich. And Greenwich Mean Time is still used by navigators, meteorologists and astronomers, who call it universal time; observatories all over the world continue to write down their findings in Universal Time. And, although Great Britain now observes Central European Time, the twenty-four-hour gate clock at Greenwich continues to show Greenwich Mean Time, which is still observed by such countries as Morocco, Ghana, Algeria and the Canary Islands.

Charles II gave lasting and world-wide import-ance to Greenwich when, 1675, he commanded Sir Christopher Wren, Knight, 'to build a small observatory within his park at Greenwich, upon the highest ground'.

From Apollo to Apollo Chapter 7

Many of our time words and the names of months
and days are memorials to changed ideas and to
gods long dead. The word *month* comes from moon
and both are believed to come from the same Indo-
European root word which means *to measure*.
The moon was the first and most convenient
measure of time. In Egyptian legend Thoth, the
moon god, is also the divider and reckoner of time.
But our months are no longer the moon-time they
once were. They do not now regularly start with
the new moon and end with its waning. They still
do in the Jewish calendar where one of the words
for a month is Hodesh, which originally simply
meant *new moon*.

When people relied so greatly on the moon to

help them keep track of time they often counted in nights rather than in days. The Roman historian Tacitus, who wrote during the first century A.D., thought it worth noting that the people of Britain reckoned time in this way, 'non dierum numerum ut nos sed noctium computant' (they do not count the number of the days, as we do, but of the nights). From this old way of counting we get the word fortnight, an abbreviation of the Old English *feowertiene niht* (fourteen nights), meaning a period of two weeks. Until the time of Shakespeare the word *sennight* (seven nights) was used as well as *week*. The witches in 'Macbeth' doom a sailor to suffer on his ship for 'weary se'nnights nine times nine'. Week probably became more frequently used and gradually drove out sennight completely; fortnight continued to be used because there was no other single word to express that period of time. Originally sennight and fortnight were written se'nnight and fort'night with an apostrophe to show that they were abbreviations, just as today we write o'clock with an apostrophe to show that it is short for *of the clock*. Perhaps eventually we shall drop that apostrophe too.

The names of the months take us back to the Romans and their many gods. Carlyle mocked the names given to the months in the French Revolutionary Calendar but they were far more logical than ours because they described the months. In this they were like old Saxon and Dutch month names. The French Revolutionaries called March *ventôse* (windy month); the old Saxons called it *hreth monath* meaning rough month, also because of its windiness.

In our calendar January, March, May and June are named after four ancient Roman gods. January, which is at the opening of the year, is named for Janus who was the god or spirit of doorways, which are 'janua' in Latin. Images of this god, with a stick in one hand, a key in the other and two heads, were set up to stand guard at many gates and entrances in Rome. With two heads Janus could see both in front and behind him at the same time; in January we are aware of the passing of the old year at the same time as we plan for the new.

March recalls Mars the patron god of Rome and the legendary father of its founders, Romulus and Remus. He is best known now as the god of war but from earliest times he was also the god of agriculture. March is traditionally the first month of spring so it was as god of agriculture that he was honoured at a time of year when plants begin to grow again after winter. Early Romans held a festival on about the fourteenth of the month, the

Left: Coin showing Janus the two-headed god of doorways. He gives his name to January. (Mansell Collection) *Below left: Mars, the patron god of Rome and the god of war and agriculture. Statue in the Museo Nazionale delle Terme, Rome.* (Mansell Collection) *Below: Juno, the goddess of marriage and the birth of children. Statue in the Museo Vaticano, Rome.* (Mansell Collection)

time of the full moon, when they drove out from the city a man who represented the Mars of the dead old year and welcomed in a younger man, as they welcomed the new spring, to represent the god in the coming year.

May is called after the goddess Maia who was responsible for the growth and increase of all living things. For a long time in many countries the first of May has been a day of open air celebrations and rejoicing at the arrival of summer. The young people of Rome used to spend it out in the fields singing and dancing. June is named after Juno, another Roman goddess. She was the wife of Jupiter, the mightiest of the gods. As she was the goddess of marriage and child-birth it was considered especially lucky to marry in her month.

April comes from the Latin *aperire* which means *to open* and is so called because it is in April that the flower and leaf buds begin to open. February is from a Roman festival of purification and sacrifice called Februa. The Christian calendar still has a festival of purification in February. It is called Candlemas Day and it is the feast of the Purification of the Virgin Mary which comes on February 2.

September, October, November, and December come from Latin words meaning seventh, eighth, ninth and tenth. They are borrowed from the very early Roman year of ten months which began in March. They are meaningless in our calendar where September, October, November and December are the ninth, tenth, eleventh and twelfth months.

The Moslem calendar is even more illogical. As it is purely lunar the months travel through the seasons. Yet some have the seasonal names which suited them at the time when the calendar was formed. Rabîa I and Rabîa II mean the first and second spring months, but they often come in winter. A more practical difficulty is that the holy month of fast, called Ramadan, sometimes comes in the hottest time of the year when going without food and drink is likely to be most dangerous to health. In Old English, September is called *haerfest* (harvest) *monath* and October was sometimes *teo monath*, simply tenth month.

Finally, July and August recall men not gods. In the early Roman calendar July was called Quintilis meaning just the fifth month of the year which began with March. In 44 B.C., shortly after Julius Cæsar had been assassinated, Mark Anthony re-named Quintilis in memory of him and his great works. Mark Anthony chose this particular month because Cæsar was born in Quintilis. Later

*The Emperor Augustus, 63 B.C.
to A.D. 14. A sardonyx cameo.*
(British Museum)

Augustus Cæsar, the first Roman emperor, could not be outdone. In 8 B.C. by order of the Senate, Sextilis, the sixth month of the year counting from March, was given his name. And to make it the same length as the month named after Julius, its 30 days were increased to 31 by taking a day from February. Sextilis had been Augustus Cæsar's lucky month, when he had begun his first consulship and had enjoyed many successes in battle.

So our month names are an illogical mixture. And other calendars are no more reasonable. An early Jewish calendar discovered at Gezer in Palestine names the months according to the agricultural work done in them: sowing, pulling flax, pruning vines and so on. By about 900 B.C. these names were dropped in favour of simple numbers. At the time of change the number of the month was sometimes put besides its old name as in the First Book of Kings, Chapter VI verse 1, which says that Solomon began to build the temple in 'the month of Ziv, which is the second month'. After about 400 years the months were given new names, this time from the Babylonian calendar where they are chiefly named for the most important festivals which take place in them. But one month, Tammuz, recalls an ancient god. Tammuz was the Babylonian god who died every year and disappeared

127

into the underworld until his love, Ishtar, went in search of him and brought him back to earth. By this story the death of vegetation in winter and its re-growth in spring were personified. The Greeks had a similar myth to explain the seasons' changes. Hades the god of the underworld saw Persephone, loved her, and carried her away to his home in the underworld. Her mother, Demeter, the goddess of fruit, crops and vegetation, went searching for her and at last was allowed to bring her back, but only for six months of every year. When she is on earth it is warm and gay with flowers; when she is in the underworld the earth suffers the miseries of winter.

The month names, then, are a reminder that once England was part of the Roman Empire. The day names are a reminder that the Saxons too invaded England. When we first adopted the seven-day week in the fourth century we kept the Roman god names for the days. Later some of them were re-named after Teutonic gods. Woden or Odin was the supreme Teutonic god. He gave his name to Wednesday (Wodensday). Like Mars he was god of war and agriculture. Woden was said to look down from his heavenly palace through a window at the world. Wednesday was once thought to be especially favourable for planting. The Romans called this

Below: Thor, the Scandinavian god of thunder, whose name lives in Thursday. A statuette in the Museum of Iceland (Gísli Gestsson) *Right: Jove, the Greek idea of a thunder god. Bust from the Museo Nazionale in Naples.* (Mansell Collection)
Below right: Venus, the Roman goddess of love, with her attendants. A silver medallion from Tarentum 1st century B.C. (British Museum)

day after Mercury.

Thursday comes from the name of Thor, who, after Odin his father, was the mightiest Teutonic deity. He was god of lightning and thunder which he made with a great hammer. Sometimes Thursday was called Thunderday. The Romans called it Jove's day (Jeudi in French); Jove was also the thunder god. Thor had power in war too. His name lives on in some place names in England, Germany and Scandinavia. There is Thorsborg in Gotland and Thorsby in Cumberland. In some countries Thursday was once a special day when people were not allowed to spin or to cut down trees.

Tuesday perpetuates the name of Tiw. He was another son of Odin, younger than Thor. All these Teutonic gods are warlike and he is especially honoured as the giver of victory. The Romans called Tuesday, Mar's day, so he had the special honour of a month and day named after him.

Freyja, the goddess of marriage, gives her name to Friday. She was the wife of Odin and the most beautiful of the goddesses. She is the Scandinavian equivalent of Venus to whom this day was dedicated in the Roman calendar.

Saturday keeps the name of the Roman god Saturn; Sunday and Monday are English translations of the Latin dies solis and dies lunae (the day of the sun and the day of the moon.) In Romance languages, like Italian and French, the day names *except* those of Saturday and Sunday come direct from the Latin. They have borrowed the Jewish Sabbath for Saturday; it becomes Sabbato in Italian and Samedi in French. These are, again, misnamed, because in Hebrew Sabbath means *day of rest,* which in Christian countries is on Sunday.

In their name for Sunday the Italians and French honour Mary the mother of Jesus. The French Dimanche and the Italian Domenica come from the Latin dies dominica meaning Lady's day. It is not surprising that these countries chose to alter the name for Sunday. It *is* surprising that the Early Christians of England did not manage to do the same. To Christians Sunday is the most important day of the week: Christ rose from the dead on a Sunday. Yet in its name we keep the reminder that long before the spread of Christianity the sun was honoured and worshipped as a great god.

Most races at some time worshipped the sun as the giver of light, heat and life, and many legends grew up to explain its journey across the sky by day and its disappearance at night. In ancient Egypt the sun god was called Ra; he was supposed to be the ancestor of all the Pharaohs. Each day he sailed across the sky in his boat. Each night he

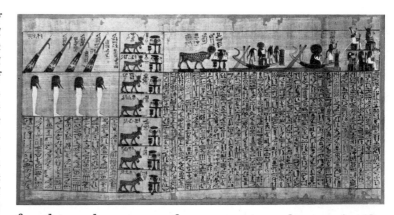

Right: Part of the Papyrus of Nekht, a royal scribe and general of Pharaoh. It shows in one boat the gods Thoth, Ra, Kheperi, and a goddess travelling in the boat of the sun. Nekht, the dead man, paddles the second boat which carries the head of Ra. From the early XIXth Dynasty. (British Museum) *Below: Helius, the Greek sun god, rising from the sea to drive his four-horse chariot across the sky. From a red figured vase* (British Museum)

fought and conquered a serpent or dragon in the underworld. The dawn was his triumphant return to the earth. He was usually depicted with the head of a hawk or sometimes a bull on a human body. Above his head a circle with a dot at its centre represented the disk of the sun. The ancient Greeks saw the sun-god Helios as driving a chariot daily across the sky. The chariot rose from the sea at dawn and plunged back into it at night; it was drawn by four white, fire-breathing horses. So at a yearly ceremony the people of Rhodes threw into the sea a chariot and four horses dedicated to the god of the sun. In later myths Apollo came to be identified with the sun; he was also the god of music and poetry and he could foresee the future. The Mexicans made human sacrifices to the sun every year; it was believed to need human blood to keep up its strength.

Clearly the worship of the sun and the moon,

especially when it required human sacrifice, was hateful to Christians. In the story of the Creation at the beginning of Genesis, the sun and moon are just called lights which God set in heaven to 'divide the day from the night' and as 'signs and for seasons and for days and years'. The editors of 'The Jerusalem Bible' which is a Roman Catholic version published in 1966 suggested that their names were deliberately omitted; 'the Sun and the Moon, deified by all the neighbouring peoples, are here no more than lamps that light the earth and regulate the calendar'.

Such words as chronometer and chronology come from the name of Chronos, the Greek god of time. Saturn was his Roman counterpart. In Greek mythology Chronos was so afraid that his children would usurp his power that he devoured all of them when they were born except Zeus who was saved by trickery. Zeus killed his father who vomited up his children. Zeus and his brothers, Poseidon and Hades, became the gods of air, water and the underworld, or the grave, which time cannot alter or destroy. Compared with Chronos, Old Father Time with his long flowing beard and his hour glass seems a gentle creature. Yet *he* carries a sickle to cut down all living things. He is often shown with a solitary lock of hair on his forehead to represent the present which can be

Goya's picture of Saturn devouring his children. In the Prado, Madrid. (Scala)

grasped and used. The rest of his head is bald to represent the lost past.

Nowadays we manage without myths but, though we seem to have time well controlled with efficient calendars and accurate clocks, there is still some mystery left. Time can play such strange tricks. Clocks say that an hour is always made up of sixty minutes but boredom can make ten minutes seem like an hour and an absorbing interest can make an hour seem like a few minutes. And time seems different to the old and the young. To children a year from one birthday to the next is like a lifetime. To their parents the years pass more quickly as they grow older.

The Mappa Mundi from Hereford Cathedral.

Philosophers, poets and scientists have tried to explain time but no one explanation will ever quite do. In the third century B.C. the Greek philosopher Aristotle put the problem in this way, 'We apprehend time only when we have marked motion' yet 'not only do we measure the movement by the time, but also the time by the movement, because they define each other.' In the fifth century A.D. St Augustine wrote, 'Time is made by the changes of things as their forms vary and are changed' so 'where there is no change there is no time'. Augustine and Aristotle seem to be using different words to express a similar idea because, like time and motion in Aristotle's definition, motion and

133

change can be used to explain each other. Motion may be defined as a change from place to place and change as a move from one condition to another.

It is hard to see how we could know time if there were no changes. The traditional place of perfect happiness for most people is a place which is out of time and therefore free from change. To the Greeks time was an endlessly turning wheel. At its centre were the ageless and unchanging gods who did not age or change because they were outside time. The heroes of Norse legend went to Valhalla after death and spent eternity joyfully feasting. The people of Mediaeval times believed that somewhere far away in the east there was an Earthly Paradise where everything was always beautiful and where death and decay were unknown. This Earthly Paradise is even shown on some thirteenth century maps as a round wall-encircled island near India. A legend of the eleventh century called 'The Navigation of St Brendan' tells how the saint searched seven years for this island and eventually found it.

At the end of the sixteenth century Sir Isaac Newton wrote of an 'absolute time' which is not liable to any change and 'of itself and from its own nature flows equally without relation to anything external'. According to this concept, even in the

Earthly Paradise where nothing changes and so presumably nothing grows, and where there are no days and nights, there still is Time. On the other hand Newton believed that motion is always relative. That is to say we can only know that we are moving when we compare or relate ourselves to something which is still. If it were possible to travel out into empty space beyond all stars and planets where there is nothing to compare ourselves with, it would be impossible to tell whether we were moving or at rest. More down to earth is the familiar sensation which we feel in a train standing at a station when we look out of the window at another one moving past ours. For a while we cannot be sure whether it is our train or the other one which is moving and we look at something we know to be still to find out what is happening to us.

Newton compared time and motion in this way: 'All motion may be accelerated or retarded but the flowing of absolute time is not liable to any change.' So, for him, time and motion were separable.

In the twentieth century, though, experiments seem to have established that there is no absolute time—that, like motion, it is relative.

In the late nineteenth century, two American physicists, Michelson and Morley, did experiments with light which later led the great mathematician Albert Einstein to conclude that the speed of light never changes, not even when the source of the light and the observer of it are moving towards or away from each other; light travels at a constant speed of 186,000 miles per second. So light behaves in a peculiar way, not at all like an ordinary, solid object.

But this peculiar behaviour of light, in always travelling at a speed of 186,000 miles per second relative to its observer, has some surprising results. Imagine two spacecraft, A and B, passing each other at extremely high speed. If the crew of A saw B out of the porthole it would seem to them to have shrunk along the direction in which it was travelling; a rectangular flag painted on B would appear as a square. If the crew of B increased the speed of their craft it would seem to shrink more and more until, if it could reach the speed of light, it would apparently shrink to nothing. And if the crew of A could see B's clock it would seem to be running slow in comparison with their own, even if the clocks were absolutely reliable and had been synchronised at the start of the flight. To the crew of B *their* clock would be keeping normal time but if *they* looked out at spacecraft A *it* would seem to them to have shrunk and to have a slow clock.

Sir Isaac Newton from a portrait by Sir Godfrey Kneller, 1702. (National Portrait Gallery)

Time seems to slow down in this way because of the peculiar behaviour of light and also because it takes an interval of time for the light waves to travel from A to B or from B to A.

Yet this time-slowing process is not just an optical illusion. Atomic particles which vibrate rather like a tuning fork can be used as tiny clocks. Experiments have shown that when these particles are moved at constant, very high, speeds their vibration slows down; the little clocks run slow.

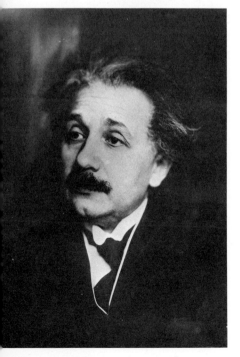

Albert Einstein, the great mathematician and physicist who first propounded the theory of relativity. (Radio Times Hulton Picture Library)

The theories of Albert Einstein, and the experiments which followed publication of them, excited people and made them wonder about the possible results of space travel. Could it, they wondered, hold the secret of perpetual youth? Scientists calculated that, if a spacecraft could be kept moving constantly at somewhere near the speed of light, one day on the spacecraft would equal ten days on earth. An astronaut might leave a newborn son on earth, travel through space for say three years, according to his time, and return to find his son a grown man of thirty years, married perhaps and with a son of his own. In three years a new father would become a grandfather.

But this could happen only if the astronaut travelled at a speed near that of light, that is at 186,000 miles a second. The American spacecraft

137

Apollo XIII even at its fastest did not cover that distance in one *hour*. Einstein's theory predicted too that time-slowing would happen only in an object travelling at a *constant* speed. There is always some acceleration and deceleration of a spacecraft, at least at take off and at landing

One undoubted conclusion from these theories seems to be that, since time can go at different rates according to the position and speed of the observer, there is no absolute time, and that, like motion, time must be relative. And since time can vary in this way it should be taken into consideration with length, breadth and width as a measurement of things. So time came to be called the fourth dimension. H. G. Wells's time traveller in 'The Time Machine' explained it to a friend in this way:

'Can an instantaneous cube exist?'

'Don't follow you', said Filby (the traveller's friend).

'Can a cube that does not last for any time at all, have a real existence?'

Filby became pensive. 'Clearly,' the time traveller proceeded, 'any real body must have extension in four directions: it must have Length, Breadth, Thickness, and—Duration.'

By very complicated reasoning, mathematicians

The Badge of the Apollo XIII flight. (NASA)

have even been able to show that, in some circumstances, past and future are different for observers moving in different directions at different speeds. One observer may say that A came before B; another that B came before A; to a third they may have happened together. We cannot say that one was right and another wrong; only that *each* was right according to his point of view. Quite apart from complicated mathematics we may see the past happening in the sky almost any night. The constellation of Orion is 300 light years away from earth; so light waves from any of its stars take 300 light years to reach us. If there was an explosion on one of its stars in January 1670 we should not have known of it until January 1970. Orion's past would be happening in our present.

Einstein's 'Special Theory of Relativity' was published in 1905 and, although anyone untrained in mathematics must find its detail extremely hard to grasp, it did make people generally think again about their concept of time. It no longer seemed possible to see it as an evenly flowing stream going at the same rate for everybody. Even their notions of past, present and future were upset. Many writers seized the opportunity to free their work from the old fixed ideas of time. We can no longer be sure that a story will follow the orderly sequence from a beginning, through a middle to an end.

J. B. Priestley is one writer who has always been fascinated by time. In his various 'Time Plays' he shows different ways of looking at it. One called 'I Have Been Here Before' attempts to explain the strange feeling we all sometimes have that we have lived through an episode in our lives already; every detail is familiar and we know what is going to happen. Dr Gortler, a character in the play, explains it like this: 'What has happened before—many times—will probably happen again. That is why some people can prophesy what is to happen. They do not see the future, as they think, but the past, what has happened before.' Priestley explains in a preface that he does not accept this theory of time though it interests him. The theory which is nearest his own is illustrated in another play, 'Time and the Conways'. Act II of the play takes us into the future of its characters through the eyes of Kay Conway, the chief character. When in Act III she returns to the time of Act I she is changed by her knowledge of what is to come. Alan, Kay's brother, explains how such things can happen. 'Now at this moment, at any moment, we're only a cross-section of our real selves. What we really are is the whole stretch of ourselves, all our time, and when we come to the end of this life, all those selves, all our time, will be *us*—the real

you, the real me. And then perhaps we'll find ourselves in another time, which is only another kind
of dream.'

The popularity of 'Dr Who' on BBC Television
showed that people were still fascinated by time
travel. But in 'Dr Who' travel in time simply
provides an excuse for a hotch-potch of unrelated
adventures in the past and in the future. Flitting
from past to future does not seem to have much
physical effect on Zoë or Dr Who. They age at the
same normal, steady rate as the characters in
'Coronation Street'.

H. G. Wells in 'The Time Machine' is one of the
few writers who attempts to show what a journey
through time would really be like. His time
traveller makes a machine which takes him forward to the year 802,701. What he finds in the
future is gloomy and depressing but the description
of the journey itself is exhilarating. As the machine·
got under way the traveller 'saw the sun hopping
swiftly across the sky, leaping it every minute and
every minute marking a day'. Then, as the machine
went faster, 'the jerking sun became a streak of
light, a brilliant arch in space' till finally, he says,
'I noted that the sun-belt swayed up and down from
solstice to solstice in a minute or less, and that
consequently my pace was over a year a minute;
and minute by minute the white snow flashed

across the world, and vanished, and was followed by the bright, brief green of spring.' It was possible to make such a journey because, the traveller explains, 'Scientific people know very well that Time is only a kind of space.'

We use time as a 'kind of space' when we measure distances by it. We talk about a place as being an hour's journey away by car and we can be understood as well as if we had said it was thirty miles away. A light year is the distance light travels in a year and we use it for convenience when we talk of the distance of stars away from the earth. We should have to use huge unimaginable numbers to express the same distance in ordinary miles.

Is time a wheel? Is time absolute? Is it relative? Is it a kind of space? It can be all these and more besides; the possibilities are bewildering. As science answers one question it seems to pose another. Only a short time ago newspapers reported that experiments in America, England and West Germany seemed to indicate that 'with certain sub-atomic particles, time not only stops but runs backwards. The "events" in their brief lives take place in reverse.' All experiments and books on time still leave us asking the same question as St Augustine in the third century A.D. 'What then *is* Time? If no one asks me, I know; if I wish to explain it to one that asketh, I know not.'

Bibliography

Achelis, Elisabeth   *The Calendar for the Modern Age,*
London 1959

Atkins, S. E. and Overall, W. H.   *Some Account of the Worshipful Company of Clockmakers of the City of London,*
London 1881

Balderston, J. L. and Squire, J. C.   *Berkeley Square,*
London 1929

Baldwin, C. E.   *The Story of the Calendar,*
Newcastle upon Tyne 1935

Britten, F. J.   *Old Clocks and Watches and their Makers,*
revised by G. H. Baillie, etc.
New York 1956

Carlyle, Thomas   *The French Revolution,*
1837 Everyman edition 1966 Volume II p. 277–279

Cipolla, Carlo M.   *Clocks and Culture,*
London 1967

Coleman, James A.   *Relativity for the Layman,*
New York 1954 Penguin Books 1964

Dunne, J. W.   *An Experiment with Time,*
London 1958

Frazer, J. G.   *The Golden Bough,*
London 1890 and many later editions

Gatty, A.   *The Book of Sundials,*
London 1889

Gélis, E.   *L'Horlogerie Ancienne,*
Paris 1949

Gould, R. T.   *The Marine Chronometer,*
London 1923

Hogg, W.   *The Book of Old Sundials and their Mottoes,*
1915

Lang, Andrew   *Myth, Ritual and Religion,*
London 1887

Laurie, P. S.   *The Old Royal Observatory,*
London 1964

Parsons, Edmund   *Time Devoured,*
London 1964

Priestley, J. B.   *Man and Time,*
London 1964

Priestley, J. B.   *Three Time Plays,*
Pan Edition 1949

Sinclair, W. M.   *Memorials of St Paul's Cathedral,*
London 1909

Spurgeon, Caroline   *Shakespeare's Imagery,*
Cambridge Paperback edition 1965

Smith, J.   *Old Scottish Clockmakers,*
Edinburgh 1921

Tait, Hugh.   *Clocks in The British Museum,*
London 1968

Ward, F. A. B.   *Time Measurement,*
London 1958

Wells, H. G.   *The Time Machine,*
London 1895 Everyman Edition 1964

Wilson, P. W.   *The Romance of the Calendar,*
London and New York 1937

The Gentleman's Magazine
1735–1752

The Horological Journal
Volume 35 1893

INDEX

RENEWALS 458-4574

DATE DUE

FEB 2 4			

GAYLORD

PRINTED IN U.S.A.